27.50
'n
phys

MHD Instabilities

MHD Instabilities   Glenn Bateman

The MIT Press
Cambridge, Massachusetts,
and London, England

Second printing, 1980
Copyright © 1978 by
The Massachusetts Institute of Technology

All rights reserved. No part of this book may be reproduced in any form or by any means, electronic or mechanical, including photocopying, recording, or by any information storage and retrieval system, without permission in writing from the publisher.

This book was set in Monophoto Times Mathematics
by Asco Trade Typesetting Limited, Hong Kong,
and printed and bound by Murray Printing Company
in the United States of America

Library of Congress Cataloging in Publication Data

Bateman, Glenn.
  MHD instabilities.

  Includes bibliographies and index.
  1. Magnetohydrodynamics—Mathematical models. 2. Plasma instabilities—Mathematical models. 3. Tokamaks—Mathematical models. I. Title.
QC718.5.M36B37   538'.6   78-17183
ISBN 0-262-02131-5

To Colleen, Timothy, and Peter

Contents

Preface xi

1

Introduction 3
1.1 Historical Perspective 4
1.2 Tokamaks 8
1.3 Outline of the Book 21
1.4 Bibliography 23
1.5 References 24

2

The MHD Equations 27
2.1 Introduction 27
2.2 Magnetic Flux and Faraday's Law 29
2.3 Motion of Magnetic Field Lines 32
2.4 The $\mathbf{J} \times \mathbf{B}$ Force 36
2.5 Conservative Forms of the MHD Equations 37
2.6 Effects Neglected by the MHD Model 40
2.7 Summary 44
2.8 References 45

3

The Rayleigh-Taylor Instability 47
3.1 Incompressible Hydrodynamic Model 48
3.2 Compressible Hydrodynamic Models 51
3.3 MHD Incompressible Rayleigh-Taylor Instability 54
3.4 Summary 57
3.5 References 58

4

MHD Equilibrium 59
4.1 Force Balance Equations 59
4.2 Surface Quantities 60
4.3 The q-Value 64
4.4 The Grad-Shafranov Equation 66
4.5 Cylinder with Elongated Cross Section—An Example of Bifurcation 71
4.6 Plasma Squeezed Between Conducting Walls 76
4.7 Tokamak Equilibrium 79
4.8 Summary 86
4.9 References 86

5

Linearized Equations and the Energy Principle 89
5.1 Linearized Equations 90
5.2 ξ-Form of the Equations 91
5.3 The Energy Principle 93
5.4 Different Forms of the Energy Principle 95
5.5 Methods Used in Linear Stability Analysis 98
5.6 Summary 99
5.7 References 100

6

Circular Cylinder Instabilities 102
6.1 Equilibrium 103
6.2 Physical Picture of Current Driven Instabilities 104
6.3 The 1-D Eigenvalue Equation 109

6.4 The 1-D Energy Principle   112
6.5 Fixed-Boundary Instabilities   114
6.6 Free-Boundary Instabilities   121
6.7 Summary   123
6.8 References   123

7

Toroidal Instabilities   125

7.1 Flux Coordinates   125
7.2 Mercier Stability Criterion   130
7.3 Applications of the Mercier Criterion   139
7.4 Large-Scale Instabilities in Axisymmetric Tori   142
7.5 Summary   146
7.6 References   146

8

High Beta Tokamaks   149

8.1 Elongated Cross Section   150
8.2 High Beta Instabilities: Surface Current Model   154
8.3 Flux Conserving Tokamaks   163
8.4 Summary   171
8.5 References   173

9

Nonlinear Instability Theory   175

9.1 Nonlinear Methods   176
9.2 Free-Boundary Instabilities   179
9.3 Fixed-Boundary Instabilities   183
9.4 References   188

10

Resistive Instabilities   190

10.1 Magnetic Islands   192
10.2 Growth of the Resistive Tearing Mode   198
10.3 $m=1$ Resistive Tearing Mode   205
10.4 Resistive Interchange Mode   209

Contents

10.5 Summary   213
10.6 References   214

11
Comparison Between Theory and Experiment   217
11.1 Sawtooth Oscillations 218
11.2 Mirnov Oscillations   223
11.3 The Disruptive Instability   226
11.4 References   230

Appendix A: Comments on the Questions   233
Appendix B: Glossary   253
Index   257

Preface

Early in 1969, Soviet physicist L. A. Artsimovich gave a series of lectures in the United States to describe the advantages of the *tokamak* device in controlled thermonuclear fusion research. Tokamaks are one kind of experimental device built in an effort to confine a hot hydrogen gas, or plasma, long enough for nuclear reactions to produce a useful amount of energy. Later in 1969, a team of British physicists reported measurements that independently confirmed the Soviet measurements on tokamaks, and within a few years more than a dozen tokamaks were constructed in laboratories around the world. By 1976, temperatures hotter than the center of the sun were achieved in tokamaks built in Oak Ridge and near Paris, confinement times of a twentieth of a second were achieved in Princeton and Moscow, and the densities needed for a reactor were produced at MIT. A strong program is underway to increase the energy density and improve confinement in order to realize the potential of controlled thermonuclear fusion as a new source of energy.

One feature common to tokamaks and most other plasma confinement devices is that their range of operation is limited by instabilities. Tokamaks exhibit an unstable behavior which mixes up the center of the plasma and produces saturated helical structures near the edge of the plasma even during routine operation. They can be subject to violent instabilities which abruptly terminate the plasma discharge and

Preface

may damage the walls when an attempt is made to operate with higher current or plasma density. For these reasons it is important to be able to predict the properties of large-scale instabilities in confined plasmas.

Many features of these instabilities are predicted by the magnetohydrodynamic (MHD) model. This simple mathematical model treats the plasma as a perfectly conducting fluid acted upon by magnetic and pressure-driven forces. It serves as a useful point of departure for predicting equilibria and large-scale instabilities in a wide variety of magnetic confinement configurations. It has been used to treat more complicated geometries than any other plasma model, and it provides a large measure of our intuition on a number of phenomena in hot plasma experiments as well as in astrophysical observations.

This book grew out of a series of lectures first presented at Oak Ridge National Laboratory in 1975. Over a period of six months scientists and engineers with a wide variety of backgrounds met once a week during the lunch hour, entirely on their own initiative. Whenever possible, I invited colleagues to give lectures in areas where they were more qualified than I. A lecture on toroidal equilibrium was presented by Lee Berry; David Nelson discussed the mathematically sufficient conditions for stability; Carl Copenhaver gave an introduction to the research on resistive instabilities; Julian Dunlap outlined the experimental observation of large-scale instabilities in tokamaks; and Fred Marcus provided a broad survey of experimental plasma confinement devices. I gratefully acknowledge here, and in specific places throughout the book, the strong influence these lectures had on this book.

About a year later I received invitations from Dieter Sigmar to present these lectures at MIT and from George Miley to present the lectures at the University of Illinois. These lectures gave me the opportunity to shift the emphasis of the material toward more practical results and more intuitive arguments. In each series of lectures, the students expressed the greatest interest in the problems, paradoxes, and open questions that appear in nearly every section of this book. The students often responded with new points of view and questions of their own which have helped me immensely.

During the subsequent rewriting of the book, new insights and simplified derivations were provided by David Nelson, John Johnson, Dieter Sigmar, Al Mense, Lee Berry, Bruce Waddell, James Rome, and many of my colleagues at Oak Ridge and around the world. Illustrations were generously contributed by many physicists, as individually acknowledged, and the line drawings were made by Eugene L. Watkin.

MHD Instabilities

1
Introduction

The objective of the controlled thermonuclear fusion program is to heat a gas composed of light elements to a temperature considerably hotter than the center of the sun and to confine this hot plasma long enough for the resulting nuclear reactions to produce more energy than was consumed. If a mixture of deuterium and tritium is used, for example, the required temperature is at least 5 to 10 keV (where 1 keV = 10^3 eV, 1 eV = 11,600 °K = 1.602×10^{-19} joules, and 13.6 eV is the ionization potential of hydrogen), and the product of particle density and confinement time must be at least

$$n\tau \geq 10^{14} \text{cm}^{-3} \text{sec} \tag{1.0.1}$$

for a useful amount of energy to be produced (Lawson, 1958; Rose and Clark, 1961). Other fuels require even higher temperatures and longer confinement times.

Large-scale instabilities have been a major obstacle in the way of progress toward controlled thermonuclear fusion. These instabilities have imposed severe limitations on the amount of current and pressure that can be confined by a magnetic field. In order to establish this perspective and give the reader a physically intuitive feeling for the subject, the experimental observations of large-scale instabilities in magnetically confined toroidal plasmas will be described in this first chapter. They will be described in roughly the order in which they were observed, with no attempt, at this point, to categorize them according to any theoretical scheme. The rest of the book will then be devoted to the prediction of these instabilities using the simplest available model—the magnetohydrodynamic (MHD) model. The particular emphasis will be on that class of confinement devices called tokamaks, described below. A more detailed description of the experimental observation of instabilities in tokamaks with some comparison between theory and experiment will be given in chapter 11.

A detailed outline of the book can be found in section 1.2. An annotated bibliography of selected references is provided at the end of each chapter. The questions scattered throughout the text are intended to stimulate the reader's interest and enlarge the scope of the book. In most cases they can be answered with less than a page of derivation, although some are quite challenging. Answers and hints are provided in the appendix. Please refer to the glossary for definitions of the specialized terms used in this field. Except as noted, the International System of units (mks) will be used thoughout this book. All electrical currents and charges will be explicit so that only the vacuum magnetic permeability and electrical permittivity are needed. For convenience the subscript 0 will be omitted so that μ_0 and ε_0 will be denoted

$\mu \equiv 4\pi \times 10^{-7}$ henrys/meter (volt·seconds/(ampere·meter))
$\varepsilon \equiv 8.854 \times 10^{-12}$ farads/meter (coulombs/(volt·meter)).

1.1 HISTORICAL PERSPECTIVE

In the first experimental attempts to produce the conditions needed for controlled thermonuclear fusion during the early 1950s, a large current was driven through a column of ionized gas in an effort to heat and confine it (Cousins and Ware, 1951; Kurchatov et al., 1956; Bishop, 1958). The current produces a magnetic field around the plasma column which thermally insulates the hot plasma from the wall and exerts an inward radial force which confines the plasma pressure. The current heats the plasma in at least two ways. Ohmic heating results from the interaction between the current and the internal resistance of the plasma. In addition, when the magnetic field is induced rapidly enough, it produces a shock wave which propagates in towards the center of the plasma column, sweeping up charged particles in its path and imparting its momentum to the ions which then become thermalized in the hot dense core of the plasma. These processes of Ohmic heating and shock heating are still widely used today.

Introduction

Question 1.1.1

The ratio of plasma thermodynamic pressure to magnetic pressure is

$$\beta \equiv 4.03 \times 10^{-16} \frac{n[\text{cm}^{-3}] T[\text{keV}]}{(B[\text{tesla}])^2} \quad (1.1.1)$$

where n is the total number of particles (ions and electrons) per cubic centimeter. The present goal for controlled thermonuclear fusion is to achieve a temperature of 10 keV and a density of 10^{14} cm^{-3} confined for one second. How many atmospheres of pressure is this? If the magnetic pressure were equal to the thermodynamic pressure ($\beta = 1$), how large would the magnetic field have to be on the plasma surface? Does this pressure ultimately have to be absorbed by the surrounding structure?

The main problem encountered in these early experiments is that the plasma column would spontaneously pinch itself off in a process that came to be known as the $m=0$ *sausage instability*, illustrated in fig. 1.1a. The mechanism of this instability is very simple. If there is a perturbation that makes parts of the plasma column constrict slightly while other parts bulge out, the longitudinal current flowing through the smaller cross-sectional area of the constricted parts produces a stronger poloidal magnetic field there, while the field around the bulged parts becomes weaker. The stronger field around each constricted part exerts a greater inward force on the plasma so that the column there constricts further. The subsequent rapidly changing magnetic field at each constriction induces a large longitudinal electric field which accelerates ions within the plasma up to energies of hundreds of keV. This process is analyzed in a beautifully written experimental paper by Anderson et al. (1958). The resulting neutron burst is short-lived and the instability rapidly throws the plasma against the wall where the plasma temperature quenches. Apparently, superthermal ions accelerated by similar instabilities led to the premature announcement of success in controlled thermonuclear fusion (Nature *181*, 217–233 (1958)). However, it is now generally accepted that such violent instabilities must be suppressed before useful thermonuclear energy can be produced.

The sausage instability can be easily stabilized by permeating the plasma column with a longitudinal magnetic field comparable in strength to that of the poloidal field around the column. One way to consider this stabilization mechanism is by noting that any constriction

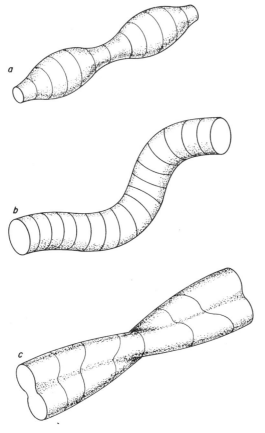

Fig. 1.1
Schematic illustrations of (a) an $m=0$ sausage instability, (b) an $m=1$ kink instability, and (c) an $m=2$ instability in a straight circular cylindrical plasma. Cross-sectional cuts are shown, not magnetic field lines.

of the plasma column requires energy to compress the longitudinal magnetic field within the plasma. It will be seen in chapter 2 that a magnetic field in a conducting fluid may be thought of as exerting pressure perpendicular to the field lines and tension along the field lines. Compressing the longitudinal field within the plasma then results in a restoring force from the increased magnetic pressure.

However, even with a moderate longitudinal magnetic field, a new instability arises in which the plasma column twists into a helical shape like a corkscrew, as shown in fig. 1.1b. The driving mechanism for this $m=1$ *kink instability* becomes clear by noting that when any part of the plasma column is bent, the poloidal field on the inner edge of the bend becomes stronger than the field on the outer edge. The resulting magnetic pressure then pushes the column to bend further and rapidly drives the plasma toward the wall. Several other ways of looking at this instability are described in chapter 6. There are some excellent photographs of this instability in the proceedings of the Second United Nations Conference on the Peaceful Uses of Atomic Energy held in Geneva in 1958. This conference marks the beginning of most of the published literature on controlled thermonuclear fusion when the subject was declassified.

In order to stabilize this $m = 1$ kink mode, the longitudinal magnetic field must be made strong enough and the plasma column fat enough so that no part of the magnetic field between the plasma and the wall closes upon itself once along the length of the plasma column. This is known as the Kruskal-Shafranov stability criterion. Roughly speaking, the tension associated with the longitudinal magnetic field prevents the plasma column from bending into a kink. A more detailed description of this stabilizing mechanism will be given in section 6.1. For a given longitudinal magnetic field, the Kruskal-Shafranov condition imposes a severe limitation on the current that can be driven through the plasma column.

Question 1.1.2

In a toroidal plasma with minor radius r, major radius R (typically $R/r \simeq 3$), toroidal field B_{tor} (long way around), and poloidal field B_{pol} (short way around) at the edge of the plasma, the Kruskal-Shafranov criterion implies

$$q \simeq \frac{rB_{tor}}{RB_{pol}} > 1. \tag{1.1.2}$$

(The meaning of the q-value will be discussed in chapter 4.) Typically, this q-value must be as high as 3 in order to avoid excessive internal instability which degrades the plasma confinement. Suppose the plasma described in question 1.1.1 were confined by the poloidal field alone— with a vacuum toroidal magnetic field present only for stability. How large does B_{tor} have to be? How large a magnetic pressure does B_{tor} exert on the coils which produce it? Is it feasible to make this field?

During the 1960s, attention turned to the poor confinement of plasmas in toroidal devices. Since the confinement was observed to be much worse than could be accounted for by simple particle collision models, it was believed that fine scale fluctuations transported plasma

Introduction

across the magnetic field. Theoretical research on MHD instabilities during this time concentrated on localized instabilities, using elegant mathematical methods to search for the most stable plasma configurations. Many of the experiments constructed during this time used elaborate magnetic field geometries in an attempt to control the exact shape and profile of the plasma column. The most successful experiments, however, involved a series of devices called *tokamaks*, built at the Kurchatov Institute in Moscow, which used the simple direct approach of a fat Ohmically-heated plasma in a strong toroidal magnetic field. Since tokamaks are now at the center of the controlled thermonuclear fusion effort at many laboratories around the world, and since they are the main subject of this book, it would be worthwhile to describe tokamaks and their instabilities in more detail in the next section.

1.2 TOKAMAKS

The essential ingredients needed for the tokamak series of experiments were known by 1958 (Dolgov-Saveliev et al., 1958), but the results were unimpressive at that time. In a series of devices built at the Kurchatov Institute during the 1960s, the size, power, and pulse length were gradually increased, almost unnoticed by the rest of the world. The turning point came when Western physicists inspected the Soviet tokamaks after the 1968 Novosibirsk IAEA Conference, and in 1969 L. A. Artsimovich gave a series of lectures in the United States. Later in that year, a team of British physicists (Peacock et al., 1969) reported bringing a laser scattering diagnostic experiment to the Kurchatov Institute and confirming Soviet claims of high temperature and comparatively long energy confinement times. Within a year the model C stellarator in Princeton was converted into a tokamak and subsequently tokamaks were built in laboratories around the world. During the 1970s, an impressive array of diagnostics were used to study tokamak plasmas and a wide range of variations have been explored—especially supplemental heating methods. Plasmas with ion temperatures of up to 2 keV have been produced with density on the order of 10^{14} particles/cm^3 and an energy confinement time on the order of 20 msec. Higher densities and longer confinement times have been reported under different conditions.

A tokamak is essentially a transformer that drives a current through a fat axisymmetric toroidal plasma in a strong magnetic field (see fig. 1.2). The word "tokamak" is a Russian acronym for "maximum

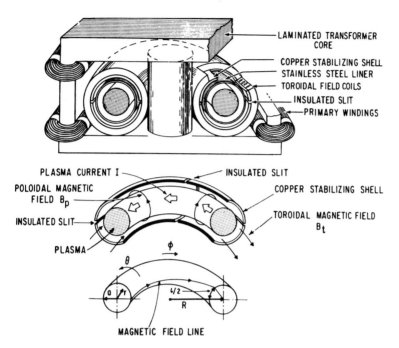

Fig. 1.2
Schematic illustration of a tokamak. From WASH-1295, courtesy of the U.S. Department of Energy.

current." The plasma is generally surrounded by two walls. The first wall is usually composed of thin stainless steel with ceramic gaps or a continuous highly-resistive bellows construction so that most of the current flows through the plasma and not through the wall. It was found that stainless steel introduces fewer impurities into the plasma than ceramic, glass, or quartz. A limiter made of a refractory material such as molybdenum or tungsten is generally used to stop down the aperture of the plasma and protect the first wall from runaway electrons which, in low density discharges, can easily burn holes through refractory materials. Since impurities play a large role in determining the characteristics of tokamak discharges, a mystique has arisen concerning the preparation of the first wall. For example, before each series of experiments on a tokamak, the first wall is "discharge cleaned" either by running a highly unstable plasma at reduced power and toroidal field to scour the wall or by running very low temperature discharges to clean the walls gently without breaking up the molecular contaminants such as CH_4 and H_2O.

In addition to the vacuum wall, most tokamaks have had a thick copper or aluminum shell that helps to center the plasma column. It will be seen in chapter 4 that the plasma column tends to expand along the major radius as a result of forces due to the plasma pressure, diamagnetism, and poloidal field. With a conducting shell present, any shift in the plasma position induces image currents and a restoring force. The plasma essentially leans against the conducting shell. In addition, there must be poloidal or vertical field coils (the coils themselves run the long way around the torus, concentric with the plasma column) to help center the plasma as the image currents in the conducting shell decay away. In some tokamaks the conducting shell is completely replaced by poloidal field coils coupled to a responsive feedback system. When a conducting shell is used, it must have insulated slits in it to allow flux and electric fields to penetrate to the plasma.

Question 1.2.1
If the magnetic flux is changed at a given rate through an open loop of wire, the voltage across the gap in the wire is minus the rate of change of flux through the loop. Suppose the flux is changed through a closed loop of resistive wire. What is the voltage between any two points on the loop?

Question 1.2.2
In a tokamak all the electric field from the transformer is concentrated at the gaps in the conducting shell. How is the electric field distributed around the toroidal plasma? Can the plasma remain axisymmetric in spite of the gaps in the shell?

The coils that produce the strong toroidal magnetic field needed for plasma stability are built around the whole assembly of plasma, walls, and poloidal field coils in many tokamaks. These toroidal field coils are often the most difficult part of the construction. It is to our advantage to make the toroidal field as high as possible and to make the aspect ratio $A = R/a$ as small as possible—a compact fat toroidal plasma—in order to drive the largest possible toroidal current through the plasma without violating the Kruskal-Shafranov condition. (There may be other advantages to low aspect ratio and high field as well.) The poloidal field coils are most effective when they are close to the plasma, but linking the two coil systems makes it very difficult to construct and repair the device. Also, the toroidal field coils must be spaced close together around the torus in order to minimize field ripple, but this leaves very little access for diagnostics, neutral beam injection,

Introduction

and adequate structural support. Each tokamak design is a compromise between these conflicting requirements.

Since the plasma discharge is usually maintained for many energy confinement times, the profiles of temperature, current density, particle density, and so forth are determined by the balance between heating and particle influx and transport effects which, incidentally, are still not well understood. In many ways, the plasma is allowed to choose its own shape and profile. Since the Kruskal-Shafranov and runaway electron limits keep the toroidal current too low for Ohmic heating alone to reach the 5 to 10 keV temperatures needed for controlled thermonuclear fusion, there has been a great deal of effort to explore other methods of heating the plasma to supplement Ohmic heating. The injection of high energy neutral beams, which pass through the magnetic field and then become ionized and trapped within the plasma, has already proved successful in doubling the ion temperature up to 2 keV at the time of this writing.

There are at least five types of large scale instabilities observed in tokamaks. We have already seen how the $m=0$ sausage and the $m=1$ kink instabilities are suppressed by a strong toroidal magnetic field. Even with this strong field, three more types of unstable behavior are observed in tokamaks: the disruptive instability, Mirnov oscillations, and sawtooth oscillations. These will be described here in roughly the order in which they were discovered.

Disruptive Instability

The *disruptive instability*, which was observed at least as early as 1963 (Gorbunov and Razumova, 1963; Artsimovich, Mirnov, and Strelkov, 1964), is an abrupt and generally unpredictable expansion of the plasma column accompanied by a large negative voltage spike kicking back against the transformer. The expansion occurs within a few hundred microseconds, after an apparently normal plasma evolution for tens or hundreds of milliseconds. Mild disruptions may occur repeatedly in a given plasma discharge while a major disruption generally terminates the discharge. As we shall see, it is possible to avoid disruptions altogether.

A disruptive instability is the generic name for an apparently complicated phenomenon with a long list of features. Helical distortions of the plasma column are observed to build up and lock together just before each disruption. Unfortunately, this cannot be used as a warning of impending disruption since roughly the same pattern frequently

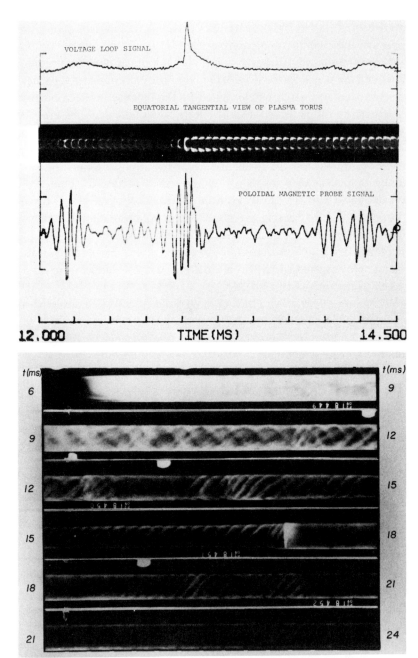

Fig. 1.3
High-speed film and streak picture showing helical structure and disruptive instabilities in the ATC tokamak at Princeton. The voltage signal is shown upside down. From R. A. Jacobsen, *Plasma Physics*, *17*, 547 (1975).

Introduction 13

occurs without disruption. Hard X rays abruptly disappear and are generally not seen again during the discharge (probably because runaway electrons are lost). There is some evidence that impurities are suddenly lost. While the electron temperature and current density profiles suddenly expand in minor radius, the plasma column suddenly shifts inward in major radius. Features even more bizarre will be discussed in chapter 11.

Two photographs taken by Jacobsen (1975) showing disruptive instabilities in the ATC tokamak are shown in fig. 1.3. Fig. 1.3a was taken by a high-speed movie camera looking tangentially through the plasma column. A fairly strong disruption can be seen at 13 msec, where the plasma expands within 100 μsec and strongly interacts with the limiter. Only the edges of the plasma are visible because the center is too hot and highly ionized to radiate much in the visible part of the spectrum. The sharp negative voltage spike (seen upside down) and the enhanced magnetic field perturbations detected at the edge of the plasma are also shown. There is a slight brightening of the plasma cross section accompanying other bursts of magnetic activity, such as the event shown here at 12.2 msec. Fig. 1.3b shows a streak picture taken by rapidly moving the film past a slit looking inward along the major radius through the plasma column or using a rotating mirror. The effective time resolution of this photograph is 15 μsec. After about 12 msec there is a clear helical structure corresponding to an $m=3$, $n=1$ mode (three lobes on any closed circuit the short way around twisted so that there is only one lobe along any closed circuit the long way around the plasma column) rotating past the slit at a frequency of 8.5 kHz. There is a strong disruption at 17.1 msec after which the helical structure is temporarily invisible (possibly because the instability has heated the formerly cold plasma at the edge of the column). It is clear that the plasma has expanded very rapidly. All disruptions shown here were mild enough for the plasma discharge to recover.

Tokamak operation has been characterized as a small island in parameter space. Consider, for example, the schematic illustration (fig. 1.4) of the bounds on toroidal current versus particle density in tokamaks in general. As the toroidal current is raised, there is a deterioration in confinement time accompanied by large helical magnetic field perturbations as the q-value drops well below 3, leading to a complete loss of plasma as the result of an $m=1$ kink instability when the q-value drops below 1 at the edge of the plasma (the Kruskal-Shafranov limit). As the plasma density is raised, the discharge becomes increasingly vulnerable to disruptive instabilities. This leads to an effective high

Introduction

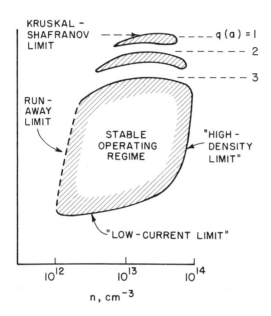

Fig. 1.4
Rough schematic of the tokamak operating regime, toroidal current vs. particle density. From WASH-1295, courtesy of the U.S. Department of Energy.

density limit to the operating parameters. This high density limit becomes lower as the toroidal current is lowered, apparently because the radial extent of the plasma column shrinks when there is not enough current to sustain the plasma heating out to the limiter. Finally, at low densities too many electrons "run away" by freely accelerating in the toroidal electric field relatively uninhibited by collisions, ultimately absorbing most of the available energy, leaving the rest of the plasma cold, and doing considerable damage when they strike the limiter or the wall. The extent of this island in parameter space depends very much on the concentration of impurities within the plasma and the condition of the wall and limiter. Disruptive instabilities can be avoided completely and optimum operating conditions achieved by operating well within this island.

It is generally agreed that the disruptive instability is the most dangerous and the most enigmatic instability observed in tokamaks. The fact that it happens so unpredictably and on such a fast time scale makes it very hard to study—both experimentally and theoretically. For this reason it is one of the least understood phenomena in tokamaks. Some additional observations and an attempt at a theoretical explanation will be offered in chapter 11.

Introduction

Question 1.2.3
As will be explained in section 4.7, the magnetic energy $(1/2\mu)\int d^3x B^2$ of a toroidal ring of plasma is roughly equivalent to the magnetic energy of a straight cylinder of plasma, with radius equal to the minor radius of the toroidal plasma, surrounded by a casing carrying the return current, with radius equal to the major radius of the toroidal plasma. Suppose the toroidal current is held fixed while the minor radius of the plasma (current profile) expands and the major radius decreases. How does the magnetic energy change?

Mirnov Oscillations

Under normal tokamak operating conditions, small oscillating helical perturbations in the poloidal magnetic field, called Mirnov oscillations, can be detected at the edge of the plasma column. These oscillations are especially prominent at the beginning of the plasma discharge when they rapidly evolve in time through a decreasing sequence of poloidal mode numbers, often starting with $m \geqslant 6$ and decreasing to $m = 2$ or 3, as shown by Mirnov and Semenov (1971b) in fig. 1.5. The helical perturbations generally have toroidal mode number $n = 1$. The perturbations are usually stronger and more coherent on the outer edge of the torus. Sometimes several harmonics are present at the same time, often locked together in phase, and sometimes the structure is not purely sinusoidal in time or space.

After running through this initial sequence, the oscillations usually settle down to a single mode number with a steady state or slowly evolving amplitude, oscillating at a frequency on the order of 10 kHz. This oscillation apparently results from a rotation of the helical structure in the direction of the electron diamagnetic drift in hot tokamaks and in the opposite direction in cold tokamaks with $T_e \lesssim 100$ eV. There is some debate over whether the direction of rotation is toroidal or poloidal—it is probably toroidal. The rate of rotation decreases when the perturbation amplitude increases, especially just before a disruption. The rotation can also be stopped at will by applying a helical perturbation with external wires. However, the amplitude usually remains unaffected by this procedure, as is seen when the externally applied perturbation is turned on and off.

When the amplitude of Mirnov oscillations becomes much larger than about one percent, the energy confinement time of the plasma drops and the total resistance increases. In general, the amplitude is larger in discharges with high current, high impurity content, and high

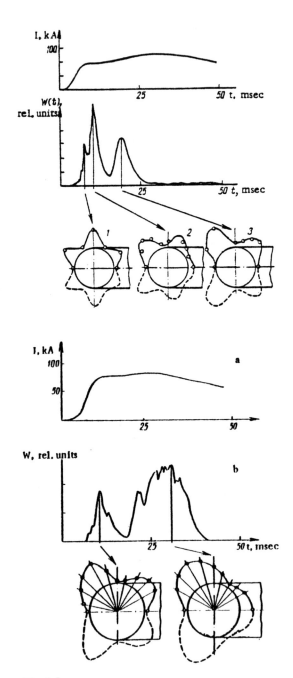

Fig. 1.5
Evolution of Mirnov oscillations in the T-3 tokamak at the Kurchatov Institute in Moscow. Oscilloscope traces show toroidal current (I) and magnetic oscillation signal strength (W). Poloidal maps show a cross section of the helical magnetic perturbation at the edge of the plasma. From S. V. Mirnov and I. B. Semenov, *Sov. Journal of Atomic Energy*, 30, 22 (1971).

Introduction

density. However, the time history of the discharge also affects the amplitude of the oscillations. For example, Mirnov and Seminov (1971a) found that the $m=2$ oscillations become prohibitively large when the toroidal current is increased so that the q-value near the edge of the plasma drops close to 2. But if a surge of current is applied in order to run quickly through the $q=2$ barrier, the oscillation amplitude remains manageably low. The amplitude of the oscillations appears to depend sensitively upon the current profile and is insensitive to the location of the conducting wall as long as the wall is not right on top of the plasma. However, since there has been no direct way to measure the current profile within hot tokamak plasmas, this profile can only be inferred from incomplete data (such as assuming a uniform toroidal electric field and Spitzer resistivity with an assumed impurity profile). The lack of adequate diagnostics and the fact that incompletely understood transport effects seem to play a large role in the slow time evolution of Mirnov oscillations has made it very difficult to make a firm link between theory and experiment.

Question 1.2.4
With sufficiently detailed measurements of the helical field perturbation outside the plasma, is it possible to infer the helical current perturbation within the plasma? For a cylindrical plasma and a current perturbation of the form $\delta(r - r_0) \sin(m\theta - kz)$, is it possible to infer r_0 from the field outside the plasma?

Sawtooth Oscillations

With high speed photographs and the detection of magnetic perturbations, it has become increasingly clear that there are internal helical structures within normal tokamak plasmas. Further indications of internal structure came from ion beam probes (Jobes, Hickock et al., 1970–1973), microwaves, and probes that could be inserted into smaller colder tokamaks. But these detection methods are limited and difficult to use routinely. A breakthrough in the field of diagnostics came when von Goeler, Stodiek, and Sautoff (1974) first used silicon surface barrier detectors to observe reproducible oscillations in the soft X rays emitted from within the plasma. This sensitive diagnostic revealed that there is a considerable level and variety of unstable activity churning around inside normal tokamak plasmas.

The term "soft X rays" generally applies to those X rays with energies between 2 and 20 keV, while "hard X rays" are those with energies

Introduction

above 100 keV. Most of the soft X rays come from electrons being captured by impurity ions (free-bound transitions), with less from impurity spectral lines (bound-bound transitions), and much less from *bremstrahlung* (free-free transitions)—especially off hydrogen. Generally, the lower energy X rays are filtered out, since they usually come from spectral line radiation that is too sensitive to the details of the ionization states and composition of the plasma. The very high energy X rays, which are mostly from runaway electrons, pass right through the detector. Since most of the X rays are emitted from the hottest and most dense part of the plasma, each detector picks up most of its signal from that part of the line of sight that passes nearest the center of the plasma. The resolution from chord to chord is typically 1 cm.

There are basically two types of X-ray signals observed in tokamaks: helical structures corresponding to Mirnov oscillations and a completely different type of signal called *sawtooth oscillations*. Extreme examples of these two types are illustrated in fig. 1.6, where the top oscilloscope trace represents the magnetic field perturbation detected

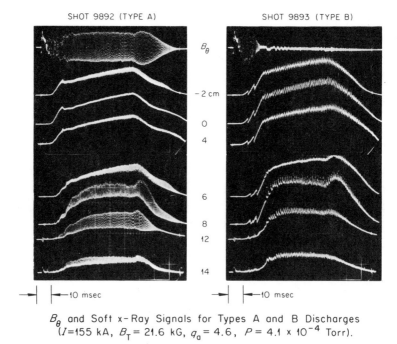

B_θ and Soft x-Ray Signals for Types A and B Discharges (I=155 kA, B_T = 21.6 kG, q_a = 4.6, P = 4.1 x 10^{-4} Torr).

Fig. 1.6
Oscilloscope traces of Mirnov oscillations (top) and the X-ray signal from different chords through the plasma for two types of discharges in ORMAK, courtesy of J. L. Dunlap, ORNL.

at the edge of the plasma, and the other seven traces represent the X-ray signal from different chords through the plasma column passing $-2, 0, 4, 6, \ldots$ cm from the magnetic axis (with the limiter at 26 cm) from the ORMAK tokamak at Oak Ridge National Laboratory schematically shown in fig. 1.7. In the case illustrated here, two strikingly different discharges were produced under nearly the same tokamak conditions. Different conditions would favor one type of discharge over the other. Since the oscilloscope traces cover the time span of the entire discharge, the first 20 msec represent the rapidly changing startup behavior, and the last 35 msec show the more gradual rundown changes, with nearly steady state conditions in between (from 20 to 70 msec).

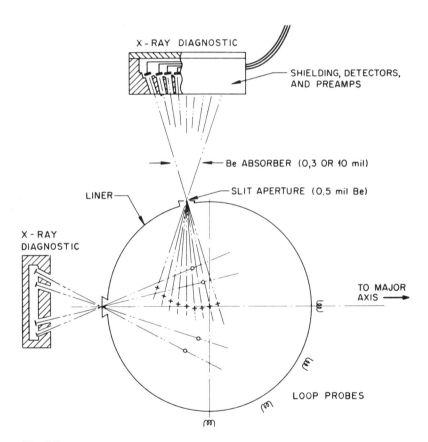

Fig. 1.7
Schematic diagram of the X-ray diagnostics and $\dot{B}$ loops on the ORMAK tokamak at Oak Ridge National Laboratory, courtesy of J. L. Dunlap, ORNL.

In the type of discharge shown on the left in fig. 1.6, there is a strong $m=2$, $n=1$ Mirnov oscillation with a corresponding X-ray oscillation at the same frequency concentrated well away from the center of the plasma. Sometimes $m \geqslant 3$, $n=1$ helical structures are observed, and there is some evidence that occasionally two or more different helical structures are observed simultaneously with maximum amplitudes at different radii. The largest amplitude $m=2$ oscillations are observed when the tokamak is operated at high density or current and especially just before a disruptive instability. Phase reversals of the $m=2$ structure are observed at two radii. Between these radii the wave form is observed to be distorted—often in the form of a sine wave truncated at the top or the bottom.

The more interesting phenomena of the sawtooth oscillations are best observed when the amplitude of the Mirnov oscillation is small,

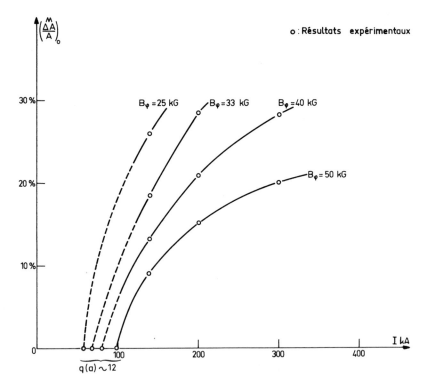

Fig. 1.8
Sawtooth oscillation strength as a function of toroidal current for several toroidal magnetic field strengths in the TFR tokamak at Fontenay-aux-Roses, France. From the TFR group, D. Launois, Seventh European Conference on Controlled Fusion and Plasma Physics (Lausanne), 2, 1 (1975).

as on the left in fig. 1.6. Sawtooth oscillations consist of an $m=0$, $n=0$ axisymmetric pulsation of the X-ray signal with an $m=1$, $n=1$ oscillation riding on top of it. Near the center of the discharge the sawteeth consist of a slow rise followed by an abrupt fall, recurring with a period on the order of one millisecond. Outside what is believed to be the $q=1$ surface, the pattern is reversed—each sawtooth has a steep rise followed by an exponential decay, with the steepness of the rise decreasing away from the $q=1$ surface. The onset of the rise in the outer sawteeth is exactly synchronized with the fall in the interior sawteeth. The period of the sawtooth oscillations increases with the amplitude of the oscillations and with increases in the plasma density. The TFR group (1975) found that the oscillations are strongly correlated with electron temperature oscillations but not with density fluctuations. Sawtooth oscillations generally become stronger as more current is driven through the plasma, as shown in fig. 1.8.

The amplitude of the $m=1$ signal riding on top of the sawtooth oscillations is strongest in the vicinity of the $q=1$ surface. The amplitude of the $m=1$ signal, however, does not seem to be directly correlated with the strength of the sawtooth oscillations; sometimes there is just a little blip of $m=1$ oscillation at the peak of the interior sawtooth, and sometimes the sawtooth is just a faint modulation of a strong $m=1$ oscillation. In general, however, the $m=1$ mode grows during the rise time of the interior sawtooth and nearly vanishes just after the fall.

1.3 OUTLINE OF THE BOOK

This book is primarily concerned with the prediction of large-scale instabilities in tokamaks by means of the simplest available mathematical model—the MHD model. A description of the major large-scale instabilities experimentally observed in tokamaks is presented in chapter 1 in an effort to give a tangible picture of the phenomena discussed in the rest of the book.

The ideal MHD model is presented in chapter 2. Here the MHD equations are described and written in a number of different ways in order to elucidate their physical meaning. The properties of a magnetic field are discussed, and it is shown that a magnetic field can be considered to be frozen into a perfectly conducting fluid so that field lines cannot break apart or change topology as long as the fluid motion is continuous. This fundamental constraint in the ideal MHD model has a strong effect on many of the phenomena predicted. Finally, since the MHD equations are often used as a first approximation under conditions where they cannot be justified, it is important for the reader to be able to estimate the important effects that have been left out. Such estimates are given for resistivity, viscosity, heat conductivity, and diffusion.

The Rayleigh-Taylor instability in a plane slab geometry is studied in chapter 3 as the simplest possible illustration of many of the techniques used to study MHD instabilities.

The formal theory of static MHD equilibrium is developed in chapter 4. Of particular importance is the idea of nested flux surfaces and quantities, such as the q-value, which are uniform over each of these surfaces. Next, the frequently used technique of reducing the vector equilibrium equations to a single partial differential equation for geometries with an ignorable coordinate is presented. Then, examples are worked out for the equilibrium of a cylindrical plasma with elongated cross section and for the vertical field needed to confine a toroidal plasma.

The mathematical formalism of the linearized MHD equations and the energy principle is developed in chapter 5. The energy principle is written in a number of different ways for later use.

Linear MHD instabilities are studied in progressively more complicated geometries in chapters 6, 7, and 8. In chapter 6, a straight circular cylindrical plasma equilibrium with periodic end conditions is considered. Essentially all the instabilities to be found in closed toroidal plasma are treated here in a simplified way. Combinations of physical models, analytic estimates, and computer solutions are used to elucidate the distinctions between pressure driven and current driven, fixed-boundary and free-boundary instabilities.

Chapter 7 on toroidal instabilities is divided into two parts. The first part contains a full derivation of the Mercier stability criterion using Hamada coordinates. A good deal of space is devoted to this derivation because it is an excellent example of the kind of analytic theory that frequently appears in the literature on linear MHD instabilities. The second part of chapter 7 reviews the computational results on large-scale instabilities in axisymmetric toroidal equilibria.

Some of the open challenges for a theoretician in controlled thermonuclear fusion are to predict the highest pressure that can be confined in a tokamak and devise ways to increase this maximum pressure. Two proposals for maximizing tokamak pressure are examined in chapter 8. In the first method the plasma cross section is elongated in an attempt to increase the toroidal current and confined pressure without violating the Kruskal-Shafranov criterion. In the second method

Introduction

auxiliary heating is used to raise the pressure more rapidly than the magnetic diffusion time scale. Then, conservation of flux forces poloidal currents to flow and help confine the plasma pressure. It is seen that both methods may be limited by a pressure driven instability, the *ballooning instability*, which is now the subject of active research.

Almost all of the material up to this point is concerned with the linear growth of instabilities that represent arbitrarily small perturbations from equilibrium. Chapter 9 surveys predictions of the nonlinear evolution and consequences of these instabilities. Most of these predictions are the result of computational work within the last few years.

The addition of a small amount of resistivity to the MHD model allows magnetic field lines to break and reconnect. This produces a whole new class of instabilities, resistive instabilities, introduced in chapter 10. There is evidence that these resistive instabilities come much closer to predicting experimental reality under tokamak conditions than do ideal MHD instabilities.

Theoretical predictions are compared to experimental reality in chapter 11. This chapter reviews the most important results in the book, relates these results to experimental observation, and suggests where new research is needed.

1.4 BIBLIOGRAPHY

If you want to learn about MHD instabilities, your reference shelf should include the collection of reprints by

Jeffrey, A. and Taniuti, T. 1966. *MHD Stability and Thermonuclear Containment*. New York: Academic Press.

An excellent review article, in much the same spirit as the present book, has been written by

Wesson, J. A. *Nuclear Fusion*. 18, 87–132 (1978).

Additional highly recommended review articles can be found in the following series of books:

Leontovich, M. A., ed. *Reviews of Plasma Physics*. English translations by the Consultants Bureau (N.Y.). See especially volumes *1* (1965); *2* (1966), *5* (1970), and *6* (1975).

A detailed presentation of theoretical methods is given by

Mercier, C. and Luc, H. 1974. *Lectures in Plasma Physics, The MHD Approach to the Problem of Plasma Confinement in Closed Magnetic Configurations*. EUR 5127e, Commission of European Communities, Luxembourg.

A good place to review both experimental and theoretical research on confined

Introduction

plasmas is the series of conferences sponsored by the International Atomic Energy Agency (IAEA) starting with the

Proceedings of the Second United Nations International Conference on the Peaceful Uses of Atomic Energy, vols. *31* and *32*, which will henceforth be referred to as the Geneva Conference (1958).

Subsequent IAEA conferences, which are published under the title *Plasma Physics and Controlled Nuclear Fusion Research*, will be referred to as the Salzburg IAEA Conf. (1961), Culham IAEA Conf. (1965), Novosibirsk IAEA Conf. (1968), Madison IAEA Conf. (1971), Tokyo IAEA Conf. (1974), and the Berchtesgaden IAEA Conf. (1976), all available from the IAEA in Vienna, Austria.

A good place to read about tokamaks is the report by the U.S. Atomic Energy Commision, WASH-1295 (1974).

Excellent advanced level reviews are written by

Artsimovich, L. A. *Nuclear Fusion.* 12, 215–252 (1972).

Furth, H. P. *Nuclear Fusion.* 15, 487–534 (1975).

The best compendium describing experimental devices for controlled fusion in laboratories around the world is

World Survey of Major Facilities in Controlled Fusion Research, published as a Nuclear Fusion Special Supplement, IAEA, Vienna, after each IAEA conference since 1969.

There are many books on plasma physics in general with some material on MHD instabilities. The ones I have used most often are

Spitzer, L. 1962. *Physics of Fully Ionized Gases.* New York: John Wiley (Interscience).

Schmidt, G. 1966. *Physics of High Temperature Plasmas.* New York: Academic Press.

Krall, N. A., and Trivelpiece, A. W. 1973. *Principles of Plasma Physics.* New York: McGraw-Hill.

Boyd, T. J. M., and Sanderson, J. J. 1969. *Plasma Dynamics.* New York: Barnes and Noble.

Rose, D. J., and Clark, M. 1961. *Plasmas and Controlled Fusion.* Cambridge: The MIT Press.

Cramer, K. R., and Pai, S-I. 1973. *Magnetofluid Dynamics for Engineers and Applied Physicists.* New York: McGraw-Hill.

1.5 REFERENCES

The first experiments for the purpose of developing controlled thermonuclear fusion were described by

S. W. Cousins and A. A. Ware, *Proc. Phys. Soc.* (London), *A64*, 159–166 (1951).

I. V. Kurchatov et al., *Sov. Journal of Atomic Energy*, *1*, 359–367 (1956).

A. Bishop, *Project Sherwood* (Reading, Mass.: Addison-Wesley, 1958).

P. C. Thonemann et al., *Nature, 181*, 217–233 (1958).

Neutron production from the $m=0$ sausage instability was proven by

O. A. Anderson, W. R. Baker, S. A. Colgate, J. Ise, and R. V. Pyle, *Phys. Rev., 110*, 1375–1387 (1958).

There are many observations of the $m=1$ kink instability at the Geneva Conference (1958) and by

G. A. Sawyer, P. L. Scott, and T. F. Stratton, *Phys. Fluids, 2*, 47–51 (1958).

A comparison between theory and experiment was made by

A. A. Ware, Salzburg IAEA Conf., Nuclear Fusion Suppl., *3*, 869–876 (1962).

Early observations of the disruptive instability were made by

E. P. Gorbunov and K. A. Razumova, *Sov. Journal of Atomic Energy, 15*, 1105–1112 (1963).

L. A. Artsimovich, S. V. Mirnov, and V. S. Strelkov, *Sov. Journal of Atomic Energy, 17*, 886–893 (1964).

For more recent work on the disruptive instability see

F. Karger et al. (Pulsator group), Berchtesgaden IAEA Conf., *1*, 267–277 (1976); Tokyo IAEA Conf., *1*, 207–213 (1974).

V. S. Vlasenkov et al., Tokyo IAEA Conf., Nuclear Fusion Suppl., 1–5 (1974).

I. H. Hutchinson, *Phys. Rev. Lett., 37*, 338–341 (1976).

T. F. R. Group. *Nuclear Fusion, 17*, 1283–1296 (1977).

For further reading on Mirnov oscillations see

S. V. Mirnov and I. B. Semenov, Madison IAEA Conf., Nuclear Fusion Suppl., 189–192 (1971); *Sov. Journal of Atomic Energy, 30*, 22–29 (1971); *Sov. Phys. JETP, 33*, 1134–1137 (1971).

J. C. Hosea, C. Bobeldijk, and D. J. Grove, Madison IAEA Conf., *2*, 425–440 (1971).

For soft X-ray observations see

S. von Goeler, Seventh European Conf. on Controlled Fusion and Plasma Physics, (Lausanne Conf.), *2*, 71–80 (1975).

TFR Group (D. Launois), Lausanne Conf., *2*, 1–13 (1975).

R. R. Smith, *Nuclear Fusion, 16*, 225–229 (1976).

S. von Goeler, W. Stodiek, and N. Sauthoff, *Phys. Rev. Lett., 33*, 1201–1203 (1974).

For other diagnostics see

R. A. Jacobsen, *Plasma Physics, 17*, 547–554 (1975).

F. C. Jobes, J. C. Hosea, and E. Hinov, Sixth European Conf. on Controlled Fusion and Plasma Physics (Moscow Conf.), *1*, 199–202 (1973).

K. Makishima et al., *Phys. Rev. Lett.*, *36*, 142–145 (1976).

For the original derivation of the Lawson criterion, see J. D. Lawson, *Proc. Phys. Soc.* (London), *B70*, 6 (1957).

2
The MHD Equations

2.1 INTRODUCTION

The abbreviation MHD stands for magnetohydrodynamics. The MHD model is one of the simplest models for describing the interaction between a perfectly conducting fluid and a magnetic field. The ideal MHD equations are

$$\rho \frac{d\mathbf{v}}{dt} = -\nabla p + \mathbf{J} \times \mathbf{B} \tag{2.1.1}$$

$$\mathbf{J} = \frac{1}{\mu} \nabla \times \mathbf{B} \tag{2.1.2}$$

$$\frac{\partial}{\partial t}\mathbf{B} = -\nabla \times \mathbf{E} \tag{2.1.3}$$

$$\mathbf{E} = -\mathbf{v} \times \mathbf{B} \tag{2.1.4}$$

$$\frac{\partial}{\partial t}p = -\mathbf{v}\cdot\nabla p - \Gamma p \nabla \cdot \mathbf{v} \tag{2.1.5}$$

$$\frac{\partial}{\partial t}\rho = -\mathbf{v}\cdot\nabla \rho - \rho \nabla \cdot \mathbf{v} \tag{2.1.6}$$

In this section a physical interpretation will be given for each term in these equations. A number of ways to rewrite and reinterpret the equations will be examined in the next four sections. And a list of the most important physical effects that have been left out of these ideal MHD equations will be given in section 6.

The state of the system at any point in space and time is given by the variables **v**, **B**, p, and ρ, where **v** is the macroscopic fluid velocity, **B** is the magnetic field, p is the thermal pressure, and ρ is the mass density. The MHD equations describe how this state advances in time. The electric field **E**, measured in the laboratory frame of reference, and the current density **J** are treated as auxiliary quantities.

The first MHD equation (2.1.1) represents the acceleration of the fluid in response to local forces. The convective derivative,

$$\frac{d}{dt} \equiv \frac{\partial}{\partial t} + \mathbf{v} \cdot \nabla, \tag{2.1.7}$$

which appears on the left in (2.1.1), represents the time rate of change at a point that follows the flow of the fluid. The pressure gradient term on the right in (2.1.1) may be thought of as the force resulting from a difference in thermal pressure on opposite sides of an infinitesimal element of fluid. The $\mathbf{J} \times \mathbf{B}$ force comes from the sum of Lorentz magnetic forces $Z_i e \mathbf{v}_i \times \mathbf{B}$ on the individual charged particles that make up the plasma ($Z_i e$ is the charge and $\mathbf{v}_i$ is the velocity of each particle). Several alternative ways of writing the $\mathbf{J} \times \mathbf{B}$ force will be examined in section 2.4. In this book the words "plasma" and "perfectly conducting fluid" will be used interchangeably.

Equation (2.1.2) is Ampere's law with the displacement current $\varepsilon \partial \mathbf{E}/\partial t$ neglected. This magnetostatic approximation is valid whenever the Alfvén velocity $v_A \equiv B/\sqrt{\mu\rho}$ is much smaller than the speed of light. It should be noted that all electrical currents are assumed to be explicit, so that throughout this book μ stands for the magnetic permeability in a vacuum

$$\mu \equiv \mu_0 = 4\pi \times 10^{-7} \text{ henrys/meter} \tag{2.1.8}$$

where the subscript 0 has been omitted for convenience.

Equation (2.1.3) is Faraday's law for the evolution of the magnetic field. A magnetic field must be divergence free

$$\nabla \cdot \mathbf{B} = 0. \tag{2.1.9}$$

If $\nabla \cdot \mathbf{B} = 0$ is used as an initial condition, Faraday's law ensures that $\nabla \cdot \mathbf{B}$ will be zero for all time. The electric field that appears in Faraday's law is the electric field in the laboratory frame of reference. When changing to a frame of reference moving with the fluid, the electric

The MHD Equations

field must be transformed by the addition of a $\mathbf{v} \times \mathbf{B}$ term. As we will see in the next section, this transformation can be derived by assuming the Galilean invariance of Faraday's law (as an approximation to its relativistic invariance). It will be shown that (2.1.4) follows from this transformation by taking the electric field to be zero in the frame of reference moving with the perfectly conducting fluid. Hence, (2.1.4) is a special form of Ohm's law. The motion of the plasma alters the magnetic field through Faraday's law (2.1.3) and Ohm's law (2.1.4), while the magnetic field acts on the motion of the plasma through the equation of motion (2.1.1).

The plasma motion alters the pressure and mass density through the thermodynamic equations (2.1.5) and (2.1.6). The $\mathbf{v} \cdot \nabla p$ and $\mathbf{v} \cdot \nabla \rho$ terms on the right side of these equations represent the effect of convection. If these terms stood alone, the pressure and density of each fluid element would never change—but would simply be carried along with the fluid. The $\Gamma p \nabla \cdot \mathbf{v}$ and $\rho \nabla \cdot \mathbf{v}$ terms in these equations represent the effect of compression and expansion. As a result of these terms, the pressure and density change as the fluid elements change size in response to changes in pressure. The constant $\Gamma = 5/3$ represents the ratio of specific heats for an ideal gas with three degrees of freedom. (The upper case Γ is used to avoid confusion with the growth rate γ later on.)

The next two sections are concerned with the consequences of Faraday's law (2.1.3) in a perfectly conducting fluid. In particular, the idea that the magnetic field lines are "frozen into the fluid" will be examined.

2.2 MAGNETIC FLUX AND FARADAY'S LAW

A magnetic field can take the form of any three-dimensional vector field that is divergence free ($\nabla \cdot \mathbf{B} = 0$). This divergence-free property implies that there are no sources or sinks of magnetic field. It does not imply that all magnetic field lines close upon themselves. A magnetic *field line* is a line in space that is everywhere tangent to the magnetic field. A field line that is not closed will generally cover a surface or fill a volume *ergodically*. This means that if we follow the field line long enough, we will eventually get arbitrarily close to any point on the surface or in the volume where the field line is ergodic.

Question 2.2.1
Consider a current-carrying ring with a straight current-carrying wire

running along its center line. Magnetic field lines form helicies around the ring. What percentage of the field lines close upon themselves and what percentage cover a toroidal surface ergodically?

Magnetic flux is the amount of magnetic field passing through any given surface

$$\psi \equiv \int d\mathbf{S} \cdot \mathbf{B}. \tag{2.2.1}$$

The divergence-free property of magnetic fields (2.1.9) implies, and is implied, by each of the following statements:
1. As much flux leaves a volume as enters it.
2. The flux through any surface spanning a given closed curve is the same.

To prove the first statement, use Gauss's theorem

$$\oint d\mathbf{S} \cdot \mathbf{B} = \int d^3x\, \mathbf{\nabla} \cdot \mathbf{B} = 0. \tag{2.2.2}$$

For the reverse implication, consider arbitrarily chosen differential volumes. For the second statement, note that there is a volume enclosed between any two surfaces spanning the same closed curve, and then use Gauss's theorem as above.

The equation $\mathbf{\nabla} \cdot \mathbf{B} = 0$ is in fact the differential form of conservation of magnetic flux. This can be seen by writing the expression $(\mathbf{\nabla} \cdot \mathbf{B}) \cdot dx \cdot dy \cdot dz$ in finite difference form within an arbitrary small rectangular box with surface areas $dx \cdot dy$, $dy \cdot dz$, and $dz \cdot dx$.

Now consider Faraday's law (2.1.3). This law comes from the following experimental observation: As the flux through any closed loop of wire is changed, the electric field observed around the wire is given by

$$\oint d\mathbf{l} \cdot \mathbf{E} = -\frac{d}{dt}\psi. \tag{2.2.3}$$

It does not matter if the flux is changing because the field is changing or because the wire is moving. Of course the wire may be replaced by any closed contour in space.

Consider a case where the field is changing and the contour is moving or deforming. In a differential interval of time, each element of the contour moves a distance $\mathbf{v} \cdot dt$, and each differential length of the contour sweeps out an area $d\mathbf{l} \times \mathbf{v} \cdot dt$. The flux through the new contour is equal to the flux through the old contour minus the flux leaving the differential area swept out by the motion of the contour. The total rate of change of flux is then given by

The MHD Equations

$$\frac{d}{dt}\psi = \int d\mathbf{S} \cdot \frac{\partial \mathbf{B}}{\partial t} - \oint d\mathbf{l} \times \mathbf{v} \cdot \mathbf{B}. \tag{2.2.4}$$

Together with (2.2.3) this implies

$$\oint d\mathbf{l} \cdot [\mathbf{E} - \mathbf{v} \times \mathbf{B}] = -\int d\mathbf{S} \cdot \frac{\partial \mathbf{B}}{\partial t}. \tag{2.2.5}$$

The electric field in (2.2.5) is the electric field moving with the contour.

The electric field around that fixed contour, which is coincident with any instantaneous position of the moving contour, is

$$\oint d\mathbf{l} \cdot \mathbf{E}_{\text{fixed}} = -\int d\mathbf{S} \cdot \frac{\partial \mathbf{B}}{\partial t}. \tag{2.2.6}$$

Here we have assumed the Galilean invariance of Faraday's law as an approximation to its relativistic invariance. That is, the magnetic field at any point in space and time is independent of the observer's frame of reference. But the electric field *is* a function of the velocity at which the observer is moving. The difference between the fixed and the moving electric field is obtained by comparing (2.2.5) and (2.2.6) for arbitrarily chosen contours

$$\mathbf{E}_{\text{fixed}} = \mathbf{E}_{\text{moving}} - \mathbf{v} \times \mathbf{B}. \tag{2.2.7}$$

In a perfectly conducting plasma the electric field is zero in the frame of reference moving with each fluid element. It then follows from (2.2.7) that the electric field observed in the laboratory frame of reference is

$$\mathbf{E}_{\text{fixed}} = -\mathbf{v} \times \mathbf{B} \tag{2.2.8}$$

where $\mathbf{v}$ is the velocity of the fluid. Hence, the ideal MHD Ohm's law (2.1.4) follows from the Galilean invariance of Faraday's law and the assumption that the electric field moving with the plasma is zero.

Note that the force acting on any moving charged particle is $Z_i e \mathbf{E}_{\text{moving}}$. Translating back to the fixed frame of reference, the force is $Z_i e \mathbf{E}_{\text{fixed}} + Z_i e \mathbf{v} \times \mathbf{B}$. The second term is the Lorentz force acting on a moving particle in a magnetic field. Summing over all the charged particles that make up the plasma, the $\mathbf{v} \times \mathbf{B}$ term results in the $\mathbf{J} \times \mathbf{B}$ force in the MHD equation of motion (2.1.1).

Question 2.2.2

A charged particle sits in a uniform magnetic field that is increasing in time. Which way does the particle move?

The MHD Equations

Question 2.2.3

The effect of the electrostatic force $\sigma \mathbf{E}_{\text{fixed}}$, where σ is the charge density of the plasma, is neglected in the MHD equation of motion (2.1.1). If (2.2.8) is multiplied by $Z_i e$ and summed over all the particles, however, it appears that $\sigma \mathbf{E}_{\text{fixed}} + \mathbf{J} \times \mathbf{B} = 0$. Does this mean that the electrostatic force cancels with the $\mathbf{J} \times \mathbf{B}$ force?

2.3 MOTION OF MAGNETIC FIELD LINES

It is the purpose of this section to explain the often quoted statement that magnetic field lines are "frozen into" any perfectly conducting fluid. The truth is that the motion of field lines is not unique. The statement that field lines move with a perfectly conducting fluid is indeed consistent with Faraday's law, but field lines could also move in other ways and still be consistent with Faraday's law. The position of field lines at any instant in time is unique, but the motion of field lines from one instant to the next is somewhat a matter of interpretation.

Fortunately there is a very simple derivation that makes all this seem reasonable, once we choose a representation for the magnetic field in the form

$$\mathbf{B} = \nabla \alpha \times \nabla \beta, \tag{2.3.1}$$

where α and β are scalar functions. The catch is that it is hard to prove that each and every magnetic field can be represented by (2.3.1). It is an easy matter to prove that (2.3.1) implies $\nabla \cdot \mathbf{B} = 0$—simply rearrange (2.3.1) to form $\mathbf{B} = \nabla \times (\alpha \nabla \beta)$ or $\mathbf{B} = -\nabla \times (\beta \nabla \alpha)$, and note that $\nabla \cdot \nabla \times \ldots = 0$. The hard part is to prove that *any* divergence-free vector field can be represented by (2.3.1). To prove this, start by imagining a pair or functions α', β' that are constant along each magnetic field line. That is, the magnetic field is everywhere tangent to the surfaces of constant α' and constant β', and these surfaces never coincide. It follows that $\nabla \alpha' \times \nabla \beta'$ is in the direction of the magnetic field at every point, and we can multiply by the appropriate function of space and time $f(\mathbf{x}, t)$ to make the magnitudes agree.

$$\mathbf{B} = f(\mathbf{x}, t) \cdot \nabla \alpha'(\mathbf{x}, t) \times \nabla \beta'(\mathbf{x}, t) \tag{2.3.2}$$

Since the magnetic field must be divergence-free,

$$\nabla \cdot \mathbf{B} = \nabla f \cdot \nabla \alpha' \times \nabla \beta' = 0,$$

it follows that $f(\mathbf{x}, t)$ must be a function of α' and β' alone, $f = f(\alpha', \beta', t)$. Then f can be absorbed into the cross product $\nabla \alpha' \times \nabla \beta'$ by defining,

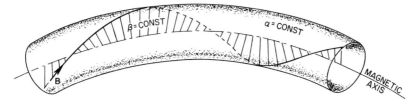

Fig. 2.1
Surfaces of constant α and β near a magnetic axis.

for example,

$$\alpha = \int_0^\alpha d\alpha' f(\alpha', \beta'), \quad \beta = \beta' \quad (2.3.3)$$

in order to derive (2.3.1).

For a concrete example, consider a magnetic field that spirals around a curve in space called the magnetic axis. The function α may be chosen to be constant on this curve and to decrease away from it so that the surfaces of constant α form tubes encircling the magnetic axis curve. Then the surfaces of constant β must be sheets twisting around the magnetic axis curve as shown in fig. 2.1. This function $\beta(\mathbf{x},t)$ is not single-valued in space; it increases around any closed path encircling the magnetic axis and, in toroidal geometry, it increases along any closed path going the long way around the toroid parallel to the magnetic axis. Thus α is single-valued and β is doubly-multivalued. It is possible to choose both α and β singly-multivalued, but apparently it is never possible to reduce this degree of multivaluedness in a toroidal configuration.

In general, α and β are not unique for any given magnetic field. For example, any function of β can be added to α, or any function of α can be added to β, without changing $\nabla\alpha \times \nabla\beta$

$$\begin{aligned}\mathbf{B} &= \nabla\alpha \times \nabla\beta \\ &= \nabla[\alpha + f(\beta)] \times \nabla\beta \\ &= \nabla\alpha \times \nabla[\beta + f(\alpha)].\end{aligned} \quad (2.3.4)$$

An equation for the time evolution of α and β can be derived by substituting $\mathbf{B} = \nabla\alpha \times \nabla\beta$ into Faraday's law for a perfectly conducting fluid

$$\frac{\partial \mathbf{B}}{\partial t} = \nabla \times (\mathbf{v} \times \mathbf{B})$$

The MHD Equations

$$\nabla \frac{\partial \alpha}{\partial t} \times \nabla \beta + \nabla \alpha \times \nabla \frac{\partial \beta}{\partial t} = \nabla \times [\mathbf{v} \times (\nabla \alpha \times \nabla \beta)]$$

$$\nabla \times \left[\frac{\partial \alpha}{\partial t} \nabla \beta - \frac{\partial \beta}{\partial t} \nabla \alpha - \mathbf{v} \cdot \nabla \beta \nabla \alpha + \mathbf{v} \cdot \nabla \alpha \nabla \beta \right] = 0$$

$$\nabla \times \left[\frac{d\alpha}{dt} \nabla \beta - \frac{d\beta}{dt} \nabla \alpha \right] = 0$$

where d/dt is the convective derivative defined by (2.1.7) for any velocity field $\mathbf{v}(\mathbf{x}, t)$. Then

$$\frac{d\alpha}{dt} \nabla \beta - \frac{d\beta}{dt} \nabla \alpha = \nabla \phi \tag{2.3.5}$$

where ϕ is an arbitrary function of space and time. The arbitrariness of ϕ reflects the fact that there is no unique way to specify how a field line evolves in time.

A perfectly valid choice for ϕ is $\phi = 0$, which leads to

$$\frac{d}{dt} \alpha = 0, \qquad \frac{d}{dt} \beta = 0. \tag{2.3.6}$$

For this particular choice of ϕ, the scalar fields $\alpha(\mathbf{x}, t)$ and $\beta(\mathbf{x}, t)$ move with the fluid. It follows that the lines of constant α and β, and therefore the magnetic field lines, must move with the fluid. Hence field lines cannot break or change topology as long as the fluid motion is continuous, in the sense that adjacent elements of fluid always remain adjacent. Since this topological invariance is a conclusion about field lines, which are unique regardless of the α, β representation, this conclusion must hold true for any choice of ϕ.

The physical interpretation of the arbitrariness of the motion of field lines is subtle. At any instant all the field lines of a magnetic field can be uniquely traced out. However, the identification of a field line at a given time with a field line at a different time is not unique. An identification can be made, and we can draw conclusions from this identification that will be valid for all other identifications as well. But we must use our intuition about field lines carefully.

Question 2.3.1
Consider a perfectly conducting rigid torus sitting between the pole pieces of a permanent magnet so that the magnetic field goes through the hole in the torus, as in fig. 2.2. Now pull the torus far away. What does the new magnetic field look like? Does it make any difference if the

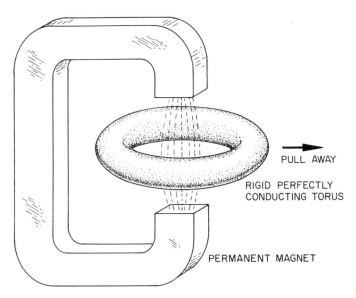

Fig. 2.2
Rigid, perfectly conducting torus being pulled from a magnetic field, for question 2.3.1.

experiment is performed in a vacuum or immersed in a perfectly conducting fluid? Does it make any difference whether there is a current flowing around the torus initially?

Question 2.3.2
Suppose a perfectly conducting fluid carrying a magnetic field encounters a nonconducting wedge with free-slip boundaries. Adjacent elements of the fluid are then separated arbitrarily far apart. Do the magnetic field lines break?

Question 2.3.3
A large, perfectly conducting, rigid pendulum bob swings into a local region of high magnetic field. What happens? Does it matter how hard the pendulum is driven into the magnetic field?

Question 2.3.4
When a magnetic dipole is rotated around its own axis, the magnetic field does not exhibit any observable rotation. How does the rotation of the ionosphere affect the earth's magnetic field?

2.4 THE J × B FORCE

The motion of a perfectly conducting fluid acts on the magnetic field through Faraday's law. The magnetic field, in turn, acts back on the fluid through the **J** × **B** force in the equation of motion (2.1.1). Various ways of writing and interpreting this **J** × **B** force will be examined in this section.

First, the **J** × **B** force can be written in terms of the curvature of the magnetic field lines and the gradient of the magnetic field strength. To do this, use (2.1.2) and vector identities to show

$$\mu \mathbf{J} \times \mathbf{B} = (\nabla \times \mathbf{B}) \times \mathbf{B} = \mathbf{B} \cdot \nabla \mathbf{B} - \tfrac{1}{2}\nabla B^2. \tag{2.4.1}$$

Then split off the unit vector $\hat{\mathbf{B}} = \mathbf{B}/|\mathbf{B}|$ along the magnetic field

$$\mu \mathbf{J} \times \mathbf{B} = B^2 \hat{\mathbf{B}} \cdot \nabla \hat{\mathbf{B}} + \tfrac{1}{2}(\hat{\mathbf{B}}\hat{\mathbf{B}} \cdot \nabla B^2 - \nabla B^2). \tag{2.4.2}$$

Note that the gradient of $\hat{\mathbf{B}}$ along the field line is just the curvature κ of the field line

$$\boldsymbol{\kappa} \equiv \hat{\mathbf{B}} \cdot \nabla \hat{\mathbf{B}} \tag{2.4.3}$$

whose magnitude is equal to the reciprocal of the radius of curvature. The other two terms in (2.4.2) can be written as the perpendicular gradient

$$\nabla_\perp \equiv \nabla - \hat{\mathbf{B}}\hat{\mathbf{B}} \cdot \nabla \tag{2.4.4}$$

of B^2. Hence

$$\mathbf{J} \times \mathbf{B} = \frac{1}{\mu}[B^2 \boldsymbol{\kappa} - \tfrac{1}{2}\nabla_\perp B^2]. \tag{2.4.5}$$

We speak of the "tension" of magnetic field lines producing a force $(1/\mu)B^2 \boldsymbol{\kappa}$ directed toward the center of curvature—in much the same way as tension acts on a string. Then the "pressure" of the magnetic field produces a force $(1/2\mu)\nabla_\perp B^2$ directed away from the region of high field strength. Bending the field produces tension and compressing it produces a restoring pressure.

Question 2.4.1

Consider a plasma in the form of a straight circular cylinder with a helical magnetic field

$$\mathbf{B} = B_\theta(r)\hat{\boldsymbol{\theta}} + B_z(r)\hat{\mathbf{z}}.$$

What are the relative contributions of magnetic pressure and magnetic

The MHD Equations

tension for this configuration? How do the relative contributions change when the longitudinal current profile $J_z(r)$ changes, if B_z is uniform and

$$B_\theta(r) = (\mu/r) \int_0^r dr\, r\, J_z(r).$$

Use the uniform current profile, for which $B_\theta = \mu r J_{z0}/2$, as a reference case.

Question 2.4.2
Suppose the magnetic field in some region all points in the same direction and has a gradient in that direction. Why is there no magnetic pressure from the parallel gradient of the magnetic field?

The $\mathbf{J} \times \mathbf{B}$ force can also be written as the divergence of the magnetic part of the Maxwell stress tensor

$$\mathbf{J} \times \mathbf{B} = \frac{1}{\mu} \nabla \cdot (\mathbf{BB} - \mathbf{I}\tfrac{1}{2}B^2) \tag{2.4.6}$$

where $\mathbf{I}$ is the unit tensor. The Cartesian components of (2.4.6) are to be interpreted as

$$(\mathbf{J} \times \mathbf{B})_i = \frac{1}{\mu} \frac{\partial}{\partial x_j}(B_i B_j - \delta_{ij}\tfrac{1}{2}B^2). \tag{2.4.7}$$

Writing the $\mathbf{J} \times \mathbf{B}$ force as the divergence of a tensor in this way is particularly useful in the conservative forms of the MHD equations.

2.5 CONSERVATIVE FORMS OF THE MHD EQUATIONS

An equation is said to be in *conservative form* when it is written as the time rate of change of a quantity set equal to the divergence of a flux. A conservative form of the ideal MHD equations in a fixed (Eulerian) coordinate system can be written

$$\frac{\partial}{\partial t}(\rho\mathbf{v}) = \nabla \cdot \left[-\rho\mathbf{vv} + \frac{1}{\mu}\mathbf{BB} - \mathbf{I}(p + B^2/2\mu) \right] \tag{2.5.1}$$

$$\frac{\partial}{\partial t}\mathbf{B} = \nabla \times (\mathbf{v} \times \mathbf{B}) = \nabla \cdot (\mathbf{vB} - \mathbf{Bv}) \tag{2.5.2}$$

$$\frac{\partial}{\partial t}\rho = -\nabla \cdot (\rho\mathbf{v}) \tag{2.5.3}$$

$$\frac{\partial}{\partial t}\left(\rho v^2 + \frac{p}{\Gamma - 1} + \frac{1}{2\mu}B^2\right)$$
kinetic + potential energy

$$= -\nabla \cdot \left[\left(\tfrac{1}{2}\rho v^2 + \frac{p}{\Gamma - 1} + p\right)\mathbf{v} + \frac{1}{\mu}\mathbf{E} \times \mathbf{B}\right]. \quad (2.5.4)$$

convection of work done Poynting
kinetic + thermo- by pressure flux
dynamic energy at a surface

These equations represent the time evolution of momentum, magnetic flux, mass density, and total energy at any point in space. Integrating each equation over any fixed volume and using Gauss's theorem (as in (2.2.2)), it can be seen that the right-hand side of each equation represents a flux through the boundaries of the volume.

Conservative forms of equations like these are especially useful in computational work. For example, finite difference forms of these equations can be devised that conserve momentum, magnetic flux, total mass, and energy exactly as the variables are advanced in time. Alternatively, conservation properties of the analytic equations can be used to test the accuracy of other, nonconservative, difference schemes.

Conservative forms of the equations are also useful for determining the natural boundary conditions which isolate the system under study from the outside world. Although the system is isolated if the net fluxes through all the boundaries are zero, it is customary to require that all the fluxes are zero through each differential element of the boundary.

For example, to conserve total mass, there must be no convection across the boundary

$$\mathbf{v}_\perp = 0. \quad (2.5.5)$$

To conserve energy, the Poynting flux must also be zero

$$(\mathbf{E} \times \mathbf{B})_\perp = 0. \quad (2.5.6)$$

To conserve magnetic flux, there must be no electric field parallel to the boundary

$$\mathbf{E}_\parallel = 0. \quad (2.5.7)$$

Condition (2.5.7) can be shown by integrating Faraday's law (2.1.3) over any fixed differential area of the boundary

$$\frac{\partial}{\partial t}\int d\mathbf{S} \cdot \mathbf{B} = -\int d\mathbf{S} \cdot \nabla \times \mathbf{E} = -\oint d\mathbf{l} \cdot \mathbf{E}. \quad (2.5.8)$$

The MHD Equations

The momentum equation is generally not used to determine a boundary condition because the force that may have to be exerted by the wall on the plasma in order to hold $v_\perp = 0$ as the system evolves is not explicitly written into the momentum equation (2.5.1).

Note that the boundary condition $E_\parallel = 0$ implies $(E \times B)_\perp = 0$. Furthermore, if the boundary is initially a flux surface, $B_\perp = 0$, and if the adjacent fluid is perfectly conducting, $E = -v \times B$, then the boundary condition $v_\perp = 0$ implies $E_\parallel = 0$, and the boundary always remains a flux surface. Under these conditions a rigid wall $v_\perp = 0$ is the only boundary condition needed. Alternatively, if the boundary is separated from the plasma by a vacuum region, or if the plasma adjacent to the boundary has zero pressure and density, then the condition $v_\perp = 0$ can be dropped and a perfectly conducting wall $E_\parallel = 0$ is all that is needed. Under all these conditions a rigid, perfectly conducting wall will completely isolate the plasma in the MHD model.

Conservative forms of the MHD equations can also be written in a Lagrangian coordinate system moving with the fluid. We have already seen that the elements of magnetic flux, α and β, can be regarded as fixed in each perfectly conducting fluid element (2.3.6). It is also possible to combine the equations for pressure (2.1.5) and mass density (2.1.6) in order to show that the *entropy*, defined by

$$e(\mathbf{x},t) \equiv p/\rho^\Gamma, \tag{2.5.9}$$

is conserved in each fluid element

$$\frac{d}{dt}\frac{p}{\rho^\Gamma} = 0. \tag{2.5.10}$$

Note that this equation does not mean that the entropy is uniform over the plasma volume.

Question 2.5.1
Do the ideal MHD equations conserve angular momentum? What is the local torque on the plasma?

Question 2.5.2
Consider a rigid disk with a pivot at the center. Attached to the disk and concentric with its center is a current-carrying ring of superconducting wire surrounded by a ring of charged pith balls. Suppose the wire is heated so that it becomes resistive and dissipates its current, causing the magnetic field to collapse. The changing magnetic field induces an electric field around the disk which exerts a force on the

charges, causing the disk to start spinning. From where did the angular momentum come?

2.6 EFFECTS NEGLECTED BY THE MHD MODEL

The MHD model is often used where it cannot be justified—to study geometries that are too complicated for the application of more sophisticated models. As such, it has provided insight and useful predictions which are then refined by adding more physical effects. In keeping with the philosophy that the MHD equations represent only a crude first approximation to reality, it would be more worthwhile to discuss some of the most important effects that have been left out of the model rather than to present a rigorous derivation of the equations with a priori estimates for their range of applicability. Good derivations of the MHD equations, with useful discussions of the assumptions involved, can be found in Boyd and Sanderson (1969), Braginskii (1965), and many other places.

The point of departure for this discussion is the moment equations —a system of equations describing the evolution of density, momentum, and energy for each species in the plasma. The most widely used reference for the moment equations is the clearly written article by Braginskii (1965). Additional useful insight can be gained from Herdan and Liley (1960), and a more detailed computation of the transport effects can be found in Shkarofsky, Johnston, and Bachynski (1966).

A fundamental assumption in the derivation of the moment equations is that all transport processes are determined locally compared to the typical scale lengths of the phenomena being studied. However in experimental devices of thermonuclear interest, the mean free path of the particles is much larger than the dimensions of the plasma— particles typically go many times around the toriodal plasma before feeling the effects of collisions. Therefore, transport effects should be determined by the global conditions of the plasma, using what is known as neoclassical transport theory (recently reviewed by Hinton and Hazeltine (1976)). As far as I know, a set of neoclassical equations suitable for studying MHD instabilities has not yet been derived, and the attempts that have been made to include the effects of long mean free paths are far beyond the scope of this book. Therefore only the effects coming from the ordinary moment equations following Braginskii (1965) will be considered here.

The state of the MHD fluid $(\mathbf{v}, p, \rho)$ can be defined in terms of the velocity moments of the distribution function $f_i(\mathbf{x}, \mathbf{v}, t)$ and the mass

The MHD Equations

m_i for each species i, including electrons. The mass density is

$$\rho \equiv \sum_i m_i n_i \tag{2.6.1}$$

where

$$n_i \equiv \int d^3v\, f_i(\mathbf{x}, \mathbf{v}, t) \tag{2.6.2}$$

is the particle density for each species. The fluid velocity is

$$\mathbf{v} \equiv \frac{1}{\rho} \sum_i m_i n_i v_i \tag{2.6.3}$$

where

$$v_i \equiv \int d^3v\, \mathbf{v}\, f_i(\mathbf{x}, \mathbf{v}, t) \tag{2.6.4}$$

is the mean velocity for each species. Finally, the plasma pressure is

$$p \equiv \sum_i n_i T_i \tag{2.6.5}$$

where

$$T_i \equiv \tfrac{1}{3} m_i \int d^3v\, (\mathbf{v} - \mathbf{v}_i)^2 f_i(\mathbf{x}, \mathbf{v}, t) \tag{2.6.6}$$

is the temperature for each species.

Equation (2.1.6) for the evolution of the mass density follows without approximation from the equations of continuity

$$\frac{\partial}{\partial t} n_i + \nabla \cdot (n_i \mathbf{v}_i) = 0. \tag{2.6.7}$$

Another equation, the equation of charge continuity,

$$\frac{\partial}{\partial t} \sigma + \nabla \cdot \mathbf{J} = 0, \tag{2.6.8}$$

where

$$\sigma \equiv \sum_i Z_i e n_i \tag{2.6.9}$$

is the charge density and

$$\mathbf{J} \equiv \sum_i Z_i e n_i \mathbf{v}_i \tag{2.6.10}$$

is the current density, is obtained by multiplying (2.6.7) by the charge $Z_i e$ for each species, including $Z = -1$ for electrons. Equations (2.6.8) to (2.6.10) are not used in MHD theory because σ does not appear anywhere else in the MHD equations.

Question 2.6.1
It follows from Ampere's law (2.1.2) that $\mathbf{V} \cdot \mathbf{J} = 0$, so that $\partial \sigma / \partial t = 0$ follows from (2.6.8). However, Maxwell's equation $\sigma = \varepsilon_0 \mathbf{V} \cdot \mathbf{E}$ together with Ohm's law (2.1.4) implies that σ is generally not constant and not zero. How can we resolve this discrepancy? Is (2.6.10) consistent with Ampere's law (2.1.2)?

The other moment equations are much more complicated. Adding the conservation of momentum equations together yields the equation of motion (2.1.1) after neglecting viscosity, any nonscalar pressure components, and electrostatic forces. The neglect of the electrostatic force $\sigma \mathbf{E}$ is a good approximation over distances much larger than the Debye length for velocities smaller than relativistic. It was pointed out by Braginskii (1965) that a pressure of 1 kg/cm^2 is equivalent to a large magnetic field of 5 kG and a huge electric field of 1.5×10^6 V/cm.

One of the simplest nonscalar pressure models is the double-adiabatic model derived by Chew, Goldberger, and Low (1956). Bernstein, Frieman, Kruskal, and Kulsrud (1958) were among the first to use this model to study MHD instabilities. They showed that if the equilibrium has isotropic pressure, the plasma will be more stable with respect to the double-adiabatic model than with respect to the usual MHD model.

Viscosity is the most important term neglected in the equation of motion. Since viscosity represents the exchange of momentum between fluid elements, the largest contribution comes from the ion component of the plasma rather than the lighter electrons. The rather complicated forms of the spatial derivative operators in the viscosity term are discussed by Braginskii (1965). The leading order contribution is written by Grimm and Johnson (1971) in the form

$$\rho \frac{d\mathbf{v}}{dt} = -\nabla p + \mathbf{J} \times \mathbf{B} - \nabla \cdot \mathbf{\Pi} \quad (2.6.11)$$

$$\mathbf{\Pi} = -3\rho v (\hat{\mathbf{B}} \hat{\mathbf{B}} - \tfrac{1}{3} \mathbf{I}) \quad (2.6.12)$$
$$\cdot [\hat{\mathbf{B}} \cdot \nabla (\hat{\mathbf{B}} \cdot \mathbf{v}) - (\hat{\mathbf{B}} \cdot \nabla \hat{\mathbf{B}}) \cdot \mathbf{v} - \tfrac{1}{3} \nabla \cdot \mathbf{v}]$$

where

$$v \simeq .96 \frac{T_i}{m_i} \tau_{ii}$$

$$\simeq \frac{6.17 \times 10^{25}}{n_e [\text{cm}^{-3}]} \frac{(T_i [\text{keV}])^{5/2}}{A^{1/2} Z^3 \Lambda} \frac{\text{cm}^2}{\text{sec}}. \quad (2.6.13)$$

The MHD Equations

Here τ_{ii} is the ion-ion collision time

$$\tau_{ii} \simeq \frac{6.7 \times 10^{10} (T_i[\text{keV}])^{3/2} A^{1/2}}{n_e[\text{cm}^{-3}] Z^3 \Lambda} \text{ sec,} \tag{2.6.14}$$

where A is the ratio of ion mass to proton mass, and Λ is the coulomb logarithm divided by 10 (see Braginskii, p. 215). There are additional contributions to the viscosity which are higher order in the collision frequency divided by the gyro-frequency

$$(\omega_{ci}\tau_{ii})^{-1} \simeq \frac{1.55 \times 10^{-19} n_e[\text{cm}^{-3}] Z A^{1/2} \Lambda}{(T_i[\text{keV}])^{3/2} B[\text{tesla}]}. \tag{2.6.15}$$

This expansion parameter is very small in plasmas of thermonuclear interest ($n_e \sim 10^{14} \text{cm}^{-3}$, $T_i > 1$ keV, $B > 2$ tesla). Note that collisional viscosity (2.6.13) rapidly gets larger with increasing temperature, which reflects the fact that there is more momentum exchange along magnetic field lines as the mean free path gets longer. This viscosity (2.6.12) is just a manifestation of $p_\parallel \neq p_\perp$. Use double-adiabatic or similar theory on time scales fast compared to τ_{ii}.

Essentially, Ohm's law comes from subtracting the momentum equations for each species. After a number of low frequency ($\omega \ll \omega_{ci}$), small Larmor radius approximations, the result is

$$\mathbf{E} = -\mathbf{v} \times \mathbf{B} + \eta \mathbf{J} + \frac{1}{n_e e} [\mathbf{J} \times \mathbf{B} - \nabla(n_e T_e)]. \tag{2.6.16}$$

The resistivity η may be thought of as a diffusion coefficient for magnetic flux

$$\frac{\partial}{\partial t}\mathbf{B} = -\nabla \times \mathbf{E} = \frac{\eta}{\mu} \nabla^2 \mathbf{B} + \cdots. \tag{2.6.17}$$

When the Spitzer resistivity is used for η, the numerical value of this diffusion coefficient η/μ is

$$\frac{\eta}{\mu} \simeq 255. Z\Lambda(T_e[\text{keV}])^{-3/2} \frac{\text{cm}^2}{\text{sec}}. \tag{2.6.18}$$

This diffusion is numerically much smaller than viscosity, but it is very important in MHD theory because it allows magnetic field lines to break and change topology.

The last term in (2.6.16) is called the Hall effect. It is the result of finite Larmor radius effects and is therefore formally smaller than the leading order effects in (2.1.16) by $(\omega_{ci}\tau_{ii})^{-1}$, given by (2.6.15). The Hall effect is believed to be responsible for the oscillatory behavior of Mirnov oscillations, but this is the subject of controversy.

Finally, there is the equation for conservation of energy, which is represented by (2.1.5) or (2.5.4) among the MHD equations. There are so many possible sources, sinks, and transfers of energy that there is no standard treatment of the subject. Numerically, the largest energy transport coefficient is parallel heat conductivity by electrons

$$\kappa_{\|e} \simeq 3.16 \frac{T_e}{m_e} \tau_{ei}$$

$$\simeq \frac{1.95 \times 10^{27} (T_e[\text{keV}])^{5/2}}{n_e[\text{cm}^{-3}] Z \Lambda} \frac{\text{cm}^2}{\text{sec}}. \qquad (2.6.19)$$

On virtually every time scale of interest, parallel heat conductivity forces the temperature to be uniform along the magnetic surfaces. By constrast, heat conductivity perpendicular to the magnetic field is smaller by $(\omega_{ce}\tau_{ei})^{-2}$, where

$$(\omega_{ce}\tau_{ei})^{-1} \simeq \frac{5.13 \times 10^{-18} n_e[\text{cm}^{-3}] Z \Lambda}{(T_e[\text{keV}])^{3/2} B[\text{tesla}]}. \qquad (2.6.20)$$

In summary then, the ideal MHD model applies to a nonexistent middle ground where collisions are frequent enough for heat conductivity and viscosity to be negligible, and collisions are infrequent enough for resistivity to be negligible. However, the applicability of the model depends strongly on the phenomena being studied. For example, for equilibrium and for those instabilities that are uniform along magnetic field lines, the effect of parallel heat conductivity and viscosity can be neglected. Incompressible instabilities, parallel and perpendicular to the magnetic field ($\mathbf{B} \cdot \nabla v = 0, \nabla_\perp \cdot \mathbf{v} = 0$), do not feel the leading order effect of viscosity at all. Away from the marginal point, ideal MHD instabilities are generally unaffected by resistivity. After the ideal MHD equations are used as a first approximation, the effect of each transport term should be tested to make successively more accurate predictions.

2.7 SUMMARY

The ideal MHD equations (2.1.1) to (2.1.6) are presented with a physical interpretation for each term in section 2.1. It is pointed out that while the electric field is zero in the frame of reference moving with the fluid, it is generally nonzero in the laboratory frame of reference, so the plasma can move perpendicular to the magnetic field.

The most important result in section 2.3 is that magnetic field lines cannot break and change topology in a perfectly conducting fluid as long as adjacent elements of the fluid always remain adjacent.

The $\mathbf{J} \times \mathbf{B}$ force can be interpreted as a combination of (1) magnetic tension due to curvature of magnetic field lines and (2) magnetic pressure due to the gradient of the field strength perpendicular to the field (2.4.4). Other interpretations of the MHD equations follow from the conservative forms of the equations (2.5.1) to (2.5.4). The boundary conditions that conserve mass, magnetic flux, and energy are $\mathbf{E}_{\|} = 0$ and $\mathbf{v}_{\perp} = 0$.

The most important effects neglected by the ideal MHD equations are (1) heat conductivity parallel to the magnetic field, (2) viscosity, and (3) resistivity. Although resistivity is numerically a small diffusion effect, it has the important consequence of allowing magnetic field topology to break.

2.8 REFERENCES

A discussion of the $\mathbf{B} = \nabla\alpha \times \nabla\beta$ representation of a magnetic field can be found in

H. Grad and H. Rubin, Geneva Conf., *31*, 190–197 (1958).

C. Mercier and H. Luc, *Lectures in Plasma Physics* (1974).

A mathematically sophisticated discussion of the motion of field lines in a perfectly conducting fluid is provided by

W. A. Newcomb, *Annals of Physics* (N.Y.), *3*, 347–385 (1958).

A discussion of the structure of magnetic fields is given by

A. I. Morozov and L. S. Solov'ev, *Reviews of Plasma Physics*, *2*, 1–102 (New York: Consultants Bureau, 1966).

For a discussion of the moment equations and other plasma models refer to

S. I. Braginskii, *Reviews of Plasma Physics*, *1*, 205–311 (New York: Consultants Bureau, 1965).

T. J. M. Boyd and J. J. Sanderson, *Plasma Dynamics* (New York: Barnes and Noble, 1969).

R. Herdan and B. S. Liley, *Reviews of Modern Physics*, *32*, 731–741 (1960).

I. P. Shkarofsky, T. W. Johnston, and M. P. Bachynski, *The Particle Kinetics of Plasmas* (Reading, Mass.: Addison-Wesley, 1966).

R. C. Grimm and J. L. Johnson, *Plasma Physics*, *14*, 617–634 (1972).

G. F. Chew, M. L. Goldberger, and F. E. Low, *Proc. Roy. Soc.* (London), *A236*, 112–118 (1956).

I. B. Bernstein, E. A. Frieman, M. D. Kruskal, and R. M. Kulsrud, *Proc. Roy. Soc.* (London), *A244*, 17–40 (1958).

F. L. Hinton and R. D. Hazeltine, *Reviews of Modern Physics*, *48*, 239–308 (1976).

3

The Rayleigh-Taylor Instability

The Rayleigh-Taylor instability occurs when a dense, incompressible fluid is supported against gravity by a fluid with less density. This instability is one of a class of instabilities that are driven by buoyancy in a statified medium. It will be seen that the *density* gradient is relevant only for incompressible fluids, while an inverted *entropy* gradient is the prerequisite for instability when an adiabatic equation of state is used, and an inverted *temperature* gradient is needed for instability when an isothermal equation of state is used. All these instabilities are related by the single physical mechanism that the density gradient in the *perturbed* state is more favorable—that is, has a lower gravitational potential energy—than the density gradient in the unperturbed state. This story will unfold when we look at a few simple examples.

There are important differences between a Rayleigh-Taylor instability driven by gravity in a plane slab and MHD instabilities in a magnetically confined plasma driven by magnetic field curvature and parallel currents as well as pressure gradients. This chapter on plane slab models is included in a book otherwise devoted to confined plasmas in order to illustrate some of the basic principles of fluid instabilities and the techniques used to study them. For this reason, the examples chosen have been simplified to their barest essentials. The first two examples, using classical ideal fluids without a magnetic field or electrical conductivity, illustrate the use of variational forms

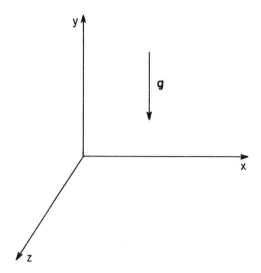

Fig. 3.1
Coordinates used in Rayleigh-Taylor instability calculation.

and the difference between incompressible, adiabatic, and isothermal models. The third example illustrates the effect of a magnetic field on a perfectly conducting MHD fluid under the same conditions. A more thorough study and review of the literature can be found in Chandrasekhar's excellent book *Hydrodynamic and Hydromagnetic Stability* (1961).

3.1 INCOMPRESSIBLE HYDRODYNAMIC MODEL

Consider an incompressible fluid with the equilibrium density stratified along the vertical (y) direction

$$\rho = \rho(y) \tag{3.1.1}$$

acted upon by a gravitational acceleration in that direction

$$\mathbf{g} = -g\hat{\mathbf{y}}. \tag{3.1.2}$$

A nonuniform density in an incompressible fluid can be obtained in practice by preparing stratified solutions of salt in water, for example. If the equilibrium is steady state ($\partial/\partial t = 0$) and stationary (no flows), the pressure gradient is given by

$$\frac{\partial p}{\partial y} = -\rho g. \tag{3.1.3}$$

The Rayleigh-Taylor Instability

This means that the pressure at any level is determined by the weight of all the fluid and the wall above.

In the absence of viscosity, surface tension, and other nonideal effects, the hydrodynamic equations are

$$\rho\left(\frac{\partial \mathbf{v}}{\partial t} + \mathbf{v}\cdot\nabla\mathbf{v}\right) = -\nabla p - \rho g \hat{\mathbf{y}} \tag{3.1.4}$$

$$\frac{\partial \rho}{\partial t} = -\mathbf{v}\cdot\nabla\rho \tag{3.1.5}$$

$$\nabla\cdot\mathbf{v} = 0. \tag{3.1.6}$$

The incompressibility condition $\nabla\cdot\mathbf{v} = 0$ means that the fluid flows into any fixed volume as fast as it leaves. This condition simplifies the equation of mass continuity (3.1.5) which now has the physical interpretation that the mass density moving with each fluid element remains unchanged.

Now start with an arbitrarily small perturbation and linearize these equations. The $\mathbf{v}\cdot\nabla\mathbf{v}$ term drops out, and the coefficients in the linearized equations are independent of time as well as coordinates x and z. It is useful to Fourier analyze the perturbation along these ignorable coordinates and then consider one Fourier harmonic at a time

$$\mathbf{v}^1(y)e^{\gamma t - i\mathbf{k}\cdot\mathbf{x}}. \tag{3.1.7}$$

The variables ρ^1, p^1, v_x^1, and v_z^1 may be eliminated by algebraic means to derive a single governing equation

$$\frac{\partial}{\partial y}\left(\rho\gamma^2\frac{\partial v_y^1}{\partial y}\right) = k^2\left(\rho\gamma^2 - g\frac{\partial \rho}{\partial y}\right)v_y^1. \tag{3.1.8}$$

Two boundary conditions must be provided for $v_y^1(y,t)$.

For any given wavenumber k there are an infinite number of solutions for $v_y^1(y)$, each with a different growth rate $\mathrm{Re}(\gamma)$ and oscillation frequency $\mathrm{Im}(\gamma)$. The set of all possible complex values of γ is called the *spectrum* of the eigenvalue problem.

Question 3.1.1

Given an arbitrary density profile $\rho(y)$, can you construct approximate solutions of (3.1.8) that are very localized relative to the scale length of the equilibrium? Do the eigenvalues γ for these solutions form a discrete set? Can the solution be localized anywhere? Based on these approximate localized solutions, where do we expect to find a point of accumulation for the spectrum?

The Rayleigh-Taylor Instability

Question 3.1.2
Construct the solution of (3.1.8) for the special case $\rho = \rho^0 e^{y/\lambda}$ with boundary conditions $v_y^1(0) = v_y^1(L) = 0$. Which are the most unstable modes?

Question 3.1.3
Find the eigenfunctions of (3.1.8) for the special case of a density profile that is uniform everywhere except for a discontinuity at $y = 0$, with boundary conditions $v_y^1(a) = v_y^1(-b) = 0$. Is this problem mathematically well posed?

We now come to a technique that is frequently used in stability analysis. The technique is to construct a variational form of the governing equation whenever it is difficult to construct even approximate solutions for it as an eigenvalue problem. Consider the situation where the boundary conditions are

$$v_y^1 = 0 \quad \text{or} \quad \partial v_y^1/\partial y = 0 \qquad (3.1.9)$$

at the upper and lower bounds of the domain. Multiply (3.1.8) by v_y^1, integrate over y, and carry out integrations by parts, using these boundary conditions to derive

$$\gamma^2 = \frac{\int dy\, g\, \partial\rho/\partial y\, v_y^2}{\int dy\, \rho\left[\frac{1}{k^2}(\partial v_y/\partial y)^2 + v_y^2\right]}. \qquad (3.1.10)$$

The advantage of (3.1.10) over (3.1.8) is that estimates for γ^2 can be made by using test functions in (3.1.10) constructed in any way compatible with the boundary conditions. The more closely the test function approximates an eigenfunction, the better the estimate for the eigenvalue γ. By using test functions with adjustable parameters, it is possible to optimize the estimate for the fastest growing instability —the largest value of γ. By using a finite subset of a complete set of test functions, it is possible to construct an eigenvalue problem whose solution involves the inversion of a finite matrix.

Question 3.1.4
Why can any test function be used to obtain a value for γ in the variational form (3.1.10) but not in the original equation (3.1.8)?

The Rayleigh-Taylor Instability 51

Question 3.1.5
Can a variational form be constructed even if (3.1.9) is not satisfied at the boundaries?

It is easy to show that there is an instability if and only if $g\,\partial\rho/\partial y$ is positive at some level in the fluid. To demonstrate this with the variational form (3.1.10), choose a test function v_y^1 that is zero everywhere except at levels where $g\,\partial\rho/\partial y$ is positive. For any such test function, γ^2 is positive but does not necessarily assume its maximum value. The fact that $\gamma = +|\gamma|$ and $\gamma = -|\gamma|$ are both solutions indicates that there is an instability for time going backward as well as for time going forward. The growth rate is independent of the magnitude of ρ but is directly related to the reciprocal of the gradient scale length $\lambda = \rho/(\partial\rho/\partial y)$. The shortest transverse wavelengths, $k^2 \to \infty$, are the most unstable, asymptotically reaching a finite point of accumulation for γ^2. If $g\,\partial\rho/\partial y$ is negative everywhere, all test functions yield negative values of γ^2, indicating that only stable oscillations exist.

Chandrasekhar (1961) shows how to construct a variational principle when a simple model of viscosity and surface tension are included. Viscosity alters the growth rate but does not stabilize. Surface tension stabilizes the shorter transverse wavelengths.

The nonlinear evolution of the Rayleigh-Taylor instability has been studied experimentally by Lewis (1950). The perturbation of the lighter fluid grows to look like elongated bubbles, while the denser fluid falls between the bubbles in thin columns or sheets. The growth of the instability makes a transition from an exponential to a linear function of time when the amplitude reaches $\sim 2.5/k$. After this transition the light fluid rises with velocity $\sim \sqrt{g/k}$, while the heavy fluid approaches free-fall.

Question 3.1.6 (Green and Niblett (1960))
In order to limit the amplitude of an exponentially growing Rayleigh-Taylor instability during the magnetic compression of a plasma, should the compression be done as slowly as possible or as quickly as possible?

3.2 COMPRESSIBLE HYDRODYNAMIC MODELS

Now let us replace the incompressibility condition $\nabla\cdot\mathbf{v} = 0$ by the adiabatic law for the fluid pressure (2.1.5). With the addition of compression, the equation of continuity for the mass density becomes

(2.1.6). The equations for pressure and density can be combined to show that the entropy

$$e \equiv p/\rho^\Gamma, \qquad \Gamma = 5/3, \tag{3.2.1}$$

of each fluid element remains constant as we follow that fluid element in the flow, as in (2.5.10).

The equilibrium for this case is the same as in section 3.1, and the same steps may be used to derive the equation governing the instability

$$\frac{\partial}{\partial y} \frac{\rho \gamma^2 \Gamma p}{\rho \gamma^2 + k^2 \Gamma p} \frac{\partial}{\partial y} v_y^1$$
$$= \left\{ \rho \gamma^2 - \frac{k^2 \rho^2 g^2}{\rho \gamma^2 + k^2 \Gamma p} - \frac{\partial}{\partial y}\left(\frac{k^2 \rho g \Gamma p}{\rho \gamma^2 + k^2 \Gamma p}\right) \right\} v_y^1. \tag{3.2.2}$$

This equation can be easily converted into a variational form, using the appropriate boundary conditions such as (3.1.9).

However, it is more instructive to simplify the governing equation by considering only transverse wavelengths much shorter than the pressure scale length λ_p,

$$k^{-2} \ll \lambda_p^2 \quad \text{where } \lambda_p \equiv 1\bigg/\left|\frac{1}{p}\frac{\partial p}{\partial y}\right|. \tag{3.2.3}$$

Then, using the equilibrium relation (3.1.3), and an estimate for the growth rate taken from the incompressible case, $\gamma^2 \lesssim g/\lambda_p$, it follows that

$$\rho \gamma^2 \ll k^2 \Gamma p. \tag{3.2.4}$$

With this approximation the governing equation reduces to

$$\frac{\gamma^2}{k^2} \frac{\partial}{\partial y}\left(\rho \frac{\partial v_y^1}{\partial y}\right) = \left\{ \rho \gamma^2 - \frac{\partial}{\partial y}(\rho g) - \frac{\rho^2 g^2}{\Gamma p} \right\} v_y^1 \tag{3.2.5}$$

and the corresponding variational form is

$$\gamma^2_{\substack{\text{adiabatic} \\ k^{-2} \ll \lambda_p^2}} = \frac{\int dy \left\{\frac{\partial}{\partial y}(\rho g) + \frac{\rho^2 g^2}{\Gamma p}\right\} v_y^2}{\int dy \rho \left\{k^{-2}\left(\frac{\partial v_y}{\partial y}\right)^2 + v_y^2\right\}}. \tag{3.2.6}$$

Why is there still an instability even when $(\partial/\partial y)\rho < 0$? Evidently the instability is not driven by an inverted density gradient. A simple model to illustrate the driving mechansim was suggested by E. G. Harris (private communication, 1975): Suppose we take a balloon from

The Rayleigh-Taylor Instability

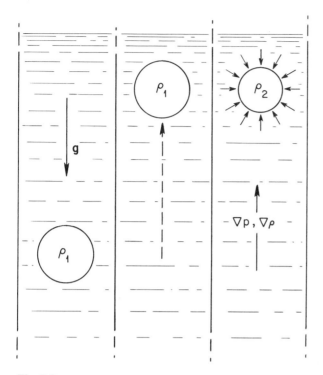

Fig. 3.2
Sequence of steps in the balloon model demonstrating the basic principles of a compressible Rayleigh-Taylor instability. Courtesy of E. G. Harris.

one level in the fluid to a higher level, as illustrated in fig. 3.2. In general, the balloon must expand or compress in order to adjust the pressure within the balloon to equal that in the ambient fluid. If, after this expansion or compression, the elevated balloon is less dense than the surrounding fluid, the resulting buoyancy will continue to drive the balloon up and, therefore, will drive an instability. The density inversion necessary for instability is present after the perturbation, but not necessarily before it.

Another way to look at the situation is to note that the variational form (3.2.6) can be rewritten

$$\gamma^2_{\substack{\text{adiabatic}\\k^{-2} \ll \lambda_p^2}} = \frac{-\int dy \frac{\rho g}{\Gamma} \frac{1}{e} \frac{\partial e}{\partial y} v_y^2}{\int dy \rho \left\{ k^{-2} \left(\frac{\partial v_y}{\partial y}\right)^2 + v_y^2 \right\}} \quad (3.2.7)$$

where e is the entropy of the fluid, defined by (3.2.1). It can be seen that the Rayleigh-Taylor instability, in this adiabatic model, rearranges the entropy of the stratified fluid. Hence the instability may be thought of as a manifestation of the "entropy wave" (which is a solution of the linearized MHD equations in an infinite uniform plasma representing a spatial discontinuity of the entropy). The single point spectrum of the entropy wave in a uniform medium ($\gamma^2 = 0$) is split into a continuous spectrum ($\gamma^2(n,k)$) in a stratified medium.

If an isothermal model were used instead of an adiabatic or incompressible model, the Rayleigh-Taylor instability would be driven by an inverted temperature gradient (colder on top, warmer at the bottom) instead of an entropy or density gradient. By an isothermal model we mean that the temperature of each fluid element does not change during the flow

$$\frac{d}{dt} T = 0, \quad \text{where } p = \rho T. \tag{3.2.8}$$

This isothermal model would follow from the adiabatic model if $\Gamma = 1$. Using this choice for the ratio of specific heats in the variational form (3.2.6) yields

$$\gamma^2_{\substack{\text{isothermal} \\ k^{-2} \ll \lambda_p^2}} = \frac{-\int dy \rho g \frac{1}{T} \frac{\partial T}{\partial y} v_y^2}{\int dy \rho \left\{ k^{-2} \left(\frac{\partial v_y}{\partial y}\right)^2 + v_y^2 \right\}}. \tag{3.2.9}$$

3.3 MHD INCOMPRESSIBLE RAYLEIGH-TAYLOR INSTABILITY

In the last two examples, a hydrodynamic model was considered in which the fluid was not electrically conducting and there was no interaction with a magnetic field. For the example in this section, an ideal MHD model will be considered in order to see the effect of a magnetic field with shear

$$\mathbf{B} = B_x(y)\hat{\mathbf{x}} + B_z(y)\hat{\mathbf{z}} \tag{3.3.1}$$

on the Rayleigh-Taylor instability in a plane slab with a vertical gravitational field (3.1.2), as illustrated in fig. 3.3. *Shear* means that the magnetic field points in different directions at different heights. In this example the magnetic field is straight and uniform at any given height, but its magnitude as well as its direction may vary with height. This

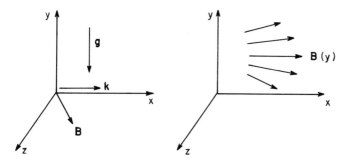

Fig. 3.3
Coordinates and sheared magnetic field for Rayleigh-Taylor instability calculation.

means that there are no magnetic curvature effects, but, in general, there is an equilibrium current in the x-z plane.

The coordinate system is rotated so that the perturbation wave vector is along the x-axis, as shown in fig. 3.3. Once again we return to an incompressible fluid model since the analysis is easier, and we have already seen how the results generalize to other thermodynamic models. The relevant equations are

$$\rho \frac{d\mathbf{v}}{dt} = -\nabla p + \mathbf{J} \times \mathbf{B} - \rho g \hat{\mathbf{y}} \tag{3.3.2}$$

$$\frac{\partial \mathbf{B}}{\partial t} = \nabla \times (\mathbf{v} \times \mathbf{B}) \tag{3.3.3}$$

together with (3.1.5) and (3.1.6).

The equilibrium is altered by the presence of a magnetic field in two ways: There must be an equilibrium current density if the magnetic field varies with height

$$\mathbf{J} = \frac{1}{\mu}\frac{\partial B_z}{\partial y}\hat{\mathbf{x}} - \frac{1}{\mu}\frac{\partial B_x}{\partial y}\hat{\mathbf{z}}, \tag{3.3.4}$$

and there is an additional $\nabla |\mathbf{B}|^2$ equilibrium force when the magnitude of the magnetic field varies with height

$$\frac{\partial}{\partial y}\left[p + \frac{1}{2\mu}(B_x^2 + B_y^2)\right] = -\rho g. \tag{3.3.5}$$

The conditions $\mathbf{v} = 0$ and $\rho = \rho(y)$ are also prescribed.

As before, the equations are linearized, only one Fourier harmonic (3.1.7) is considered at a time, and other variables are eliminated in

The Rayleigh-Taylor Instability

favor of v_y^1 in order to derive the governing equation

$$\frac{\partial}{\partial y}\left\{[\rho\gamma^2 + (\mathbf{k}\cdot\mathbf{B})^2]\frac{\partial}{\partial y}v_y^1\right\}$$

$$= k^2\left[\rho\gamma^2 + (\mathbf{k}\cdot\mathbf{B})^2 - g\frac{\partial\rho}{\partial y}\right]v_y^1 \tag{3.3.6}$$

and the corresponding variational form

$$\gamma^2 = \frac{\int dy\left\{g\frac{\partial\rho}{\partial y}v_y^2 - (\mathbf{k}\cdot\mathbf{B})^2\left[v_y^2 + k^{-2}\left(\frac{\partial v_y}{\partial y}\right)^2\right]\right\}}{\int dy\rho\left[k^{-2}\left(\frac{\partial v_y}{\partial y}\right)^2 + v_y^2\right]}. \tag{3.3.7}$$

It can be seen from this variational form that those components of the perturbation with wave number perpendicular to the magnetic field feel no effect from the field at all, while those components with wave number parallel to the field tend to be stabilized, with the shortest wavelengths stabilized the most. When the wave number is perpendicular to the magnetic field, the undulations of the instability line up along the field lines, and field lines can be interchanged without changing the magnetic energy. When the wave number is parallel to the magnetic field, the undulations bend the field lines and increase magnetic tension. The shorter the wavelength, the more the bending. For any given wave number k, shear tends to localize the instability to the level where $\mathbf{k}\cdot\mathbf{B} = 0$. In the immediate neighborhood of this level, field lines can interchange without bending very much.

Question 3.3.1
Does bending of the field lines change the magnitude of the magnetic field, and therefore the magnetic energy, even in the absence of compression?

Question 3.3.2
What is the growth rate of the instability when the equilibrium density and magnetic field are uniform except for a discontinuity, as in question 3.1.3? Is there a wall stabilization effect due to image currents?

Question 3.3.3 (Wesson (1970), Berge (1972))
An inverted pendulum (with rigid rod) provides a mechanical analogue to the Rayleigh-Taylor instability. Such a pendulum can be dynamically stabilized by shaking it up and down with a sufficiently large

The Rayleigh-Taylor Instability

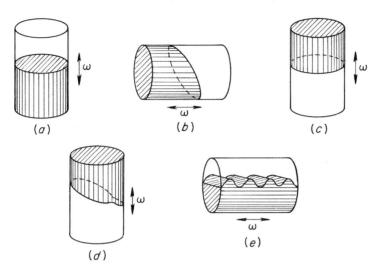

Fig. 3.4
Dynamic stabilization of a Rayleigh-Taylor instability. The dense fluid is shaded. Each configuration depends on the frequency, amplitude, and direction of the forced oscillation. Drawn from photographs of the experiment by G. H. Wolf, *Z. Physik*, **227**, 291 (1969).

frequency and amplitude. When the oscillation amplitude a is much smaller than the pendulum length l, $\varepsilon \equiv a/l \ll 1$, the inverted pendulum can be stabilized if the driving frequency ω_s is much larger than the natural frequency of the pendulum $\omega_0 \equiv \sqrt{g/l}$, $\omega_s \gg \sqrt{2\omega_0/\varepsilon}$. Also, a normal pendulum becomes unstable if its natural frequency is close to a half integer multiple of the driving frequency $\omega_0 \simeq (n/2)\omega_s$. Suppose we have a collection of pendulums, some inverted and some normal, whose lengths vary continuously from some minimum cutoff (where the motion is dominated by damping) to some maximum length. Is it possible to stabilize the inverted ones without destabilizing any of the normal ones?

3.4 SUMMARY

The Rayleigh-Taylor instability is driven by an inverted equilibrium density gradient (3.1.10) in an incompressible fluid model (3.1.6), by an inverted entropy gradient (3.2.7) in an adiabatic model (2.1.5), and by an inverted temperature gradient (3.2.9) in an isothermal model (3.2.8). In a perfectly conducting fluid, short wavelengths parallel to the magnetic field are suppressed (3.3.7) while wavelengths perpendicular to the field are not affected. The advantages of a variational form such as (3.1.10) over the corresponding eigenvalue equation (3.1.8) are demonstrated.

3.5 REFERENCES

The most complete and lengthy discussion of the Rayleigh-Taylor and related instabilities is given by

S. Chandrasekhar, *Hydrodynamic and Hydromagnetic Stability* (Oxford: Clarendon Press, 1961).

For more recent references, together with a treatment of the initial value problem, see

R. A. Axford, Los Alamos report LA-5378 (June, 1974).

Experimental demonstrations of the instability are discussed by

D. J. Lewis, *Proc. Roy. Soc.* (London), *A202*, 81–96 (1950).

T. S. Green and G. B. F. Niblett, *Nuclear Fusion*, *1*, 42–46 (1960).

D. Albares, N. A. Krall, and C. L. Oxley, *Phys. Fluids*, *4*, 1031–1036 (1961).

N. Rostoker, in *Plasma Physics in Theory and Application*, W. B. Kunkel, ed. (New York: McGraw-Hill, 1966).

For more on dynamic stabilization see

G. Berge, *Nuclear Fusion*, *12*, 99–117 (1972).

J. A. Wesson, *Phys. Fluids*, *13*, 761–766 (1970).

G. H. Wolf, *Zeitschrift für Physik*, *227*, 291–300 (1969).

4
MHD Equilibrium

In the MHD theory of magnetically confined plasmas, *equilibrium* means the complete balance of forces. Since an instability is the tendency to move away from equilibrium, it is appropriate to begin the study of instabilities with a review of the important properties of the equilibria to be considered. The first three sections of this chapter are concerned with the general conditions for static, steady state, scalar pressure MHD equilibria with arbitrary shape. The discussion of the "q-value," defined in section 4.3, is of particular importance. The Grad-Shafranov equation for axisymmetric toroidal equilibria is then derived in the next section. Finally, three examples are worked out in the last three sections in order to illustrate some special features, as well as specific problems, encountered in equilibrium calculations.

4.1 FORCE BALANCE EQUATIONS

The standard MHD equilibrium equations are

$$\nabla p = \mathbf{J} \times \mathbf{B} \tag{4.1.1}$$

$$\mathbf{J} = \frac{1}{\mu} \nabla \times \mathbf{B} \tag{4.1.2}$$

$$\nabla \cdot \mathbf{B} = 0 \tag{4.1.3}$$

MHD Equilibrium

These equations apply to plasmas with scalar pressure, in steady state ($\partial/\partial t = 0$), with no flow ($\mathbf{v} = 0$), and no body forces such as gravity or neutral gas pressure. These are the most commonly used assumptions for tokamaks and pinches, where they are usually well justified.

There are a number of other ways to write these equilibrium equations. When the $\mathbf{J} \times \mathbf{B}$ force is written as the sum of magnetic pressure and magnetic tension, as in (2.4.5), the equations become

$$\nabla_\perp (p + B^2/2\mu) = \frac{1}{\mu} B^2 \boldsymbol{\kappa} \tag{4.1.4}$$

where $\boldsymbol{\kappa} = \hat{\mathbf{B}} \cdot \nabla \hat{\mathbf{B}}$ is the curvature of the field lines. Alternatively, from (2.4.1) we have

$$\nabla (p + B^2/2\mu) = \frac{1}{\mu} \mathbf{B} \cdot \nabla \mathbf{B}. \tag{4.1.5}$$

The stress tensor form of $\mathbf{J} \times \mathbf{B}$, (2.4.6) or (2.4.7), produces

$$\nabla \cdot \left[\frac{1}{\mu} \mathbf{BB} - \mathbf{I}(p + B^2/2\mu) \right] = 0. \tag{4.1.6}$$

Integrating this divergence of the stress tensor over an arbitrary volume and using Gauss's theorem provides an integral form of the equilibrium equations

$$\oint d\mathbf{S} \cdot [(p + B^2/2\mu)\mathbf{I} - \mathbf{BB}] = 0 \tag{4.1.7}$$

where S is any closed surface with normal $\hat{\mathbf{n}}$.

Question 4.1.1
Can we take any magnetic field and find a suitable MHD equilibrium for it? Can you find a magnetic field for which $\mathbf{J} \times \mathbf{B}$ is not the gradient of a scalar function?

4.2 SURFACE QUANTITIES

If a magnetic field line is followed long enough, it will either close upon itself or continue indefinitely to cover a surface, or fill a volume, or leave the bounded domain. This book is concerned with magnetically confined plasmas in which most of the field lines continue indefinitely, ergodically covering a set of simply nested toroidal surfaces. Any surface that is covered by a magnetic field line is called a *magnetic surface*. A field line is said to cover a magnetic surface *ergodically* if it passes

MHD Equilibrium

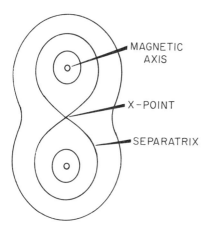

Fig. 4.1
Cross section of magnetic surfaces and two magnetic axes.

arbitrarily close to any point on the surface. A toroidal surface, or *toroid*, is any surface that is topologically equivalent to a torus (a torus has a circular hole and a circular cross section). Simply nested magnetic surfaces, one within the other, surround a field line called the *magnetic axis*. If there is more than one magnetic axis, as illustrated in fig. 4.1, the topology of the flux surfaces must change between regions containing different magnetic axes, and the surface marking this change in topology is called the *separatrix*. The easiest way to find the separatrix is to look for an *x*-point in the cross section of the magnetic surfaces.

In general, some of the magnetic field lines must close upon themselves after a finite number of transits the long way around the torus. These closed field lines lie on toroidal magnetic surfaces, called *rational surfaces*, nested in between the ergodically covered toroidal magnetic surfaces—in much the same way as rational numbers are interspersed between irrational numbers. When the field lines on a particular rational surface have the same topology as that of an instability being considered, the surface is called a *mode-rational surface*. These surfaces play an important role in the theory of instabilities, as we shall see later in the book.

There are many possibilities for magnetic fields besides the simply nested toroidal configuration. Magnetic surfaces can break up into thin filaments, *magnetic islands*, twisting through the plasma. These will be discussed at length in chapter 10. The islands themselves can carry smaller islands within them, these smaller islands can, in turn, carry

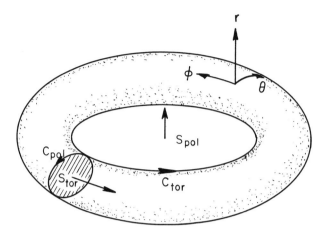

Fig. 4.2
Toroidal flux surface showing cut surfaces and contours.

successively finer and finer island structures. Alternatively, a magnetic field line can fill a volume quasi-ergodically—randomly wandering around so that it comes arbitrarily close to any point in the volume. At the other extreme, it is possible for all the field lines to be closed, leaving no well-defined flux surfaces. Each of these alternatives has been considered in the literature but will not be considered in this book.

Now consider a given toroidal magnetic surface and consider the *cut surfaces* which span across the hole in the toroid, S_{pol}, and across a cross section of the toroid, S_{tor}, as shown in fig. 4.2. Through any cross section of the toroid, S_{tor}, the *toroidal flux* is

$$\psi_{tor} \equiv \int_{S_{tor}} d\mathbf{S} \cdot \mathbf{B} = \text{flux the long way around}, \quad (4.2.1)$$

and through any cut surface spanning the center of the toroid, S_{pol}, the *poloidal flux* is

$$\psi_{pol} \equiv \int_{S_{pol}} d\mathbf{S} \cdot \mathbf{B} = \text{flux the short way around}. \quad (4.2.2)$$

As shown in section 2.2, the flux is the same for all surfaces spanning the same contour. There is no flux through the toroidal magnetic surface since the magnetic field is everywhere tangent to it. Therefore, the flux is the same through any topologically equivalent contour (C_{tor} or C_{pol}) on the flux surface. It follows that both ψ_{tor} and ψ_{pol} are surface quantities. A *surface quantity* is any variable that is uniform

MHD Equilibrium

over a magnetic surface. Conversely, it can be shown that all *flux surfaces* (surfaces of constant ψ_{tor} and ψ_{pol}) are magnetic surfaces.

There are many other surface quantities used in the theory of MHD equilibria. For example, the fact that pressure is a surface quantity follows from $\mathbf{B} \cdot \nabla p = 0$, which is a direct consequence of $\nabla p = \mathbf{J} \times \mathbf{B}$. If the pressure varies from surface to surface, so that $|\nabla p| \neq 0$ except at isolated magnetic surfaces, then the toroidal and poloidal currents are also surface quantities

$$I_{tor} \equiv \int_{S_{tor}} d\mathbf{S} \cdot \mathbf{J} = \frac{1}{\mu} \oint_{C_{pol}} d\mathbf{l} \cdot \mathbf{B} \tag{4.2.3}$$

= total current the long way around within a magnetic surface,

$$I_{pol} \equiv \int_{S_{tor}} d\mathbf{S} \cdot \mathbf{J} = \frac{1}{\mu} \oint_{C_{pol}} d\mathbf{l} \cdot \mathbf{B} \tag{4.2.4}$$

= total current through the hole in a toroidal magnetic surface.

This follows from the fact that $\mathbf{J} \cdot \nabla p = \mathbf{J} \cdot \mathbf{J} \times \mathbf{B} = 0$ implies that no current passes through the magnetic surfaces.

Question 4.2.1
If the pressure is uniform on each magnetic surface, how does a mirror machine confine a plasma?

Note that the standard notation in the more mathematically abstract literature is ψ (or Ψ) and I for the toroidal flux and current within a magnetic surface, with χ and J for the poloidal flux and current passing through a ribbon between the magnetic axis and the magnetic surface, following the notation of Kruskal and Kulsrud (1958). However, in the literature on axisymmetric toroidal configurations, ψ stands for a stream function that is proportional to the poloidal flux. I have tried to avoid all this confusion by explicitly labeling fluxes and currents either poloidal or toroidal and then retaining the stream function notation ψ when it occurs.

Also, instead of using a ribbon between the magnetic axis and the magnetic surface, I have used the surface spanning the center of the toroid to define the poloidal flux and current. The poloidal flux includes the flux in the transformer, and the poloidal current includes the current in the toroidal field coils. Explicit reference to the magnetic axis is not needed.

4.3 THE q-VALUE

The *q-value* (sometimes called the "safety factor") is defined as the number of times a magnetic field line winds the long way around the toroid divided by the number of times it winds the short way around, in the limit of an infinite number of times

$$q \equiv \lim \frac{\text{winding number long way}}{\text{winding number short way}}. \qquad (4.3.1)$$

Because field lines never cross, the q-value is the same for all field lines on a magnetic surface. Hence, the q-value is a surface quantity. As we shall see, it is a critically important quantity in the theory of MHD instabilities—often appearing in stability conditions and in expressions for growth rates.

The q-value is sometimes computed by following a magnetic field line and taking the ratio of its winding numbers. However, it is generally better to use the following alternative expression

$$q = d\psi_{\text{tor}}/d\psi_{\text{pol}} \qquad (4.3.2)$$

which is applicable to any magnetic field configuration in any closed geometry.

A formal derivation of (4.3.2), making use of a special coordinate system in which the magnetic field lines appear as straight lines, will be discussed in section 7.2. However, Kruskal ((1965), section N) devised a simple derivation of (4.3.2) which will be reproduced here.

Split open a differentially thick collection of neighboring flux surfaces and flatten them out, as shown in fig. 4.3. The surfaces labeled V_1 and V_2 are toroidal flux surfaces; the angles labeled φ and θ are arbitrary toroidal and poloidal angles. For the special case to be

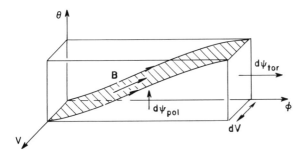

Fig. 4.3
Illustration for Kruskal's derivation of $q = d\psi_{\text{tor}}/d\psi_{\text{pol}}$.

MHD Equilibrium

considered first, all the magnetic field lines are assumed to close upon themselves once around the toroid—so that $q = 1$, with no shear. Consider a ribbon running tangent to the field lines between the flux surfaces, as shown with shading in fig. 4.3. Now deform the ribbon until it forms the right end and lower edge of the box. The differential amount of toroidal flux through the end is $+d\psi_{tor}$ and the differential amount of poloidal flux through the lower edge is $-d\psi_{pol}$. No flux passes through the sides of the box or through the original ribbon, since these surfaces are everywhere tangent to the magnetic field. All the flux entering the box through the lower edge must leave it through the end

$$d\psi_{pol} = d\psi_{tor} \quad \text{or} \quad d\psi_{tor}/d\psi_{pol} = 1 = q.$$

Now use the same argument for the case of a field line that makes n transits the long way around (along φ) for every m transits the short way around (along θ) to obtain the expression for the q-value on rational magnetic surfaces

$$\frac{d\psi_{tor}}{d\psi_{pol}} = \frac{n}{m} = q.$$

Equation (4.3.2) then follows by continuity for irrational q-values, and the differential argument still applies when shear is present, that is, when $dq(V)/dV \neq 0$.

Question 4.3.1
Suppose the poloidal component of the magnetic field vanishes at the inner edge of an axisymmetric torus, leaving only the toroidal field. The field wraps around the outer edge of the torus on the same surface with a nonzero poloidal component. Do these field lines on the same surface have different q-values?

Question 4.3.2
What is the q-value at the magnetic axis? Is it different from the q-value of the field lines in the immediate neighborhood of the magnetic axis?

The q-value is often confused with various definitions of the *helical pitch* of a magnetic field line. The helical pitch is a local quantity which may differ from place to place on a magnetic surface, while the q-value is a topological property of the field lines, being everywhere the same on a magnetic surface. The helical pitch is equal to the reciprocal of the q-value only for a straight circular cylinder configuration.

The *rotational transform*, another quantity often used in the literature, is just the reciprocal of the q-value

$$\iota = 1/q. \tag{4.3.3}$$

When expressed in radians, the rotational transform is denoted $\iota = 2\pi/q$.

4.4 THE GRAD-SHAFRANOV EQUATION

A straight cylinder, an axisymmetric toroid, and a configuration with helical symmetry, each with arbitrary cross-sectional shape, are the three types of configuration that have at least one ignorable coordinate. By introducing a stream function ψ, it is possible to reduce the equilibrium equations in each of these three cases to a single partial differential equation in one unknown. Since this reduction greatly simplifies the problem of computing the equilibrium for any given set of profiles and boundary conditions, computer programs using these reduced equilibrium equations are now standard at many laboratories around the world.

This section will be devoted to a derivation and discussion of the Grad-Shafranov equation, which is the reduced equilibrium equation for an axisymmetric toroid. A table of the reduced equations for all three configurations with continuous symmetry will be given at the end of the section.

For a derivation of the Grad-Shafranov equation, use a cylindrical coordinate system (R, y, φ) whose axis coincides with the center line of the toroid, as shown in fig. 4.4. Because the toroidal angle φ is an

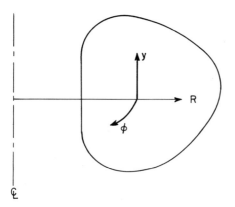

Fig. 4.4
Coordinate system used for axisymmetric toroidal equilibria.

MHD Equilibrium

ignorable coordinate ($\partial/\partial\varphi = 0$), the poloidal magnetic field may be expressed in terms of the toroidal component of the vector potential A_φ alone

$$\mathbf{B} = \nabla \times (A_\varphi \hat{\varphi}) + B_\varphi \hat{\varphi}. \tag{4.4.1}$$

Instead of using the vector potential, it is customary to use a stream function ψ defined by

$$\mathbf{B} = \nabla \times (\psi \nabla \varphi) + RB_\varphi \nabla \varphi \tag{4.4.2}$$
$$= \nabla \psi \times \nabla \varphi + B_\varphi \hat{\varphi}$$
$$= \frac{1}{R}\frac{\partial \psi}{\partial y}\hat{\mathbf{R}} - \frac{1}{R}\frac{\partial \psi}{\partial R}\hat{\mathbf{y}} + B_\varphi \hat{\varphi}$$

This stream function is proportional to the poloidal flux

$$\psi = \psi_{\text{pol}}/2\pi \tag{4.4.3}$$

as can be seen by substituting (4.4.2) into (4.2.2). Hence ψ is a surface quantity, and, in general, it may be used to label the flux surfaces.

An expression for the current density follows from (4.4.2)

$$\mu \mathbf{J} = \nabla \times \mathbf{B} = -\frac{1}{R}\Delta^*\psi \hat{\varphi} + \nabla(RB_\varphi) \times \nabla \varphi \tag{4.4.4}$$

where Δ^* is a special operator defined by

$$\Delta^*\psi \equiv R\frac{\partial}{\partial R}\frac{1}{R}\frac{\partial \psi}{\partial R} + \frac{\partial^2 \psi}{\partial y^2}. \tag{4.4.5}$$

From $\mathbf{B}\cdot\nabla p = \mathbf{B}\cdot\mathbf{J} \times \mathbf{B} = 0$ it follows that the pressure is a surface quantity so that

$$p = p(\psi). \tag{4.4.6}$$

Then by substitution into

$$\nabla p = \mathbf{J} \times \mathbf{B}$$
$$\mu p'(\psi)\nabla\psi = -R^2\Delta^*\psi\nabla\psi - RB_\varphi(1/R^2)\nabla(RB_\varphi), \tag{4.4.7}$$

it follows that $\nabla(RB_\varphi)$ must be in the direction of $\nabla\psi$. Thus RB_φ must be a surface quantity

$$RB_\varphi = I(\psi). \tag{4.4.8}$$

In fact, $I(\psi)$ is proportional to the total poloidal current

$$I(\psi) = \frac{\mu}{2\pi}\int_{S_{\text{pol}}} d\mathbf{S}\cdot\mathbf{J}_{\text{pol}} = \mu\frac{I_{\text{pol}}(\psi)}{2\pi}. \tag{4.4.9}$$

The Grad-Shafranov equation then follows from (4.4.7)

$$-\Delta^*\psi = R^2\,\mu p'(\psi) + II'(\psi) \tag{4.4.10}$$

$$\Delta^* \equiv R\frac{\partial}{\partial R}\frac{1}{R}\frac{\partial}{\partial R} + \frac{\partial^2}{\partial y^2}$$

The $\Delta^*\psi$ term on the left side of (4.4.10) represents that part of the confinement coming from the toroidal current crossed with the poloidal magnetic field (in fact $-\Delta^*\psi = RJ_\varphi$), while the II' term on the right side of (4.4.10) represents that part of the confinement from the poloidal current crossed with the toroidal magnetic field.

The standard method for computing axisymmetric tokamak equilibria is to specify the functions $p = p(\psi)$ and $I = I(\psi)$, together with boundary conditions or externally imposed constraints on ψ, and then invert the Laplacian-like operator $\Delta^*\psi$ to determine $\psi = \psi(R, y)$. When the profiles or boundary conditions are nonlinear, this process is iterated until ψ converges to a consistent solution. This procedure is somewhat awkward because the source functions $p'(\psi)$ and $II'(\psi)$ must be specified as a function of ψ, whose spatial dependence is not known until the equation is solved. In general, it is not possible to specify $\psi(R, y)$ first and then determine $p(\psi)$ and $I(\psi)$ for a toroid with finite aspect ratio (Bateman (1974)). Only an infinitesimal subset of all possible flux functions and magnetic fields are suitable for MHD equilibria.

Both toroidal and poloidal currents can contribute to plasma confinement. The degree to which the poloidal currents contribute to the confinement can be parameterized by the diamagnetism μ_J and the poloidal beta β_{pol} defined by

$$\mu_J \equiv \tfrac{1}{2}\langle I_{\text{edge}}^2 - I^2(\psi)\rangle / \overline{B}_{\text{pol}}^2 \langle R\rangle^2 \tag{4.4.11}$$

$$\beta_{\text{pol}} \equiv 1 + \mu_J \tag{4.4.12}$$

where

$$\langle\cdots\rangle \equiv \int_{S_{\text{tor}}} dS\cdots \bigg/ \int_{S_{\text{tor}}} dS \tag{4.4.13}$$

is an average over the cross section of the plasma and

$$\overline{B}_{\text{pol}} \equiv \oint_{C_{\text{pol}}} d\mathbf{l}\cdot\mathbf{B}_{\text{pol}} \bigg/ \oint_{C_{\text{pol}}} dl \tag{4.4.14}$$

MHD Equilibrium

is an average around the poloidal circumference of the plasma. These definitions are not standard (Shafranov (1971), (1974), Callen and Dory (1972)).

There are three interesting limits of the Grad-Shafranov equation. In one limit, there is no poloidal current within the plasma, $(II'(\psi) = 0)$, the toroidal magnetic field is a vacuum field which drops off like $1/R$, $(RB_\varphi = R_0 B_{\varphi 0})$, and the pressure is confined entirely by the toroidal current crossed with the poloidal magnetic field it induces. This case, with $\beta_{pol} = 1$, has been widely studied. In the second extreme case, the current density is entirely parallel to the magnetic field, the plasma pressure is zero or at least uniform, $(\nabla p = p'(\psi) = 0)$, and there is no magnetic confinement at all. The Grad-Shafranov equation becomes $-\Delta^*\psi = II'(\psi)$. Any region of the plasma where this condition applies is called a *force-free* region. If it applies over the whole region, then $\beta_{pol} = 0$. Finally, in the opposite extreme case, the plasma pressure is confined almost entirely by the poloidal current crossed with the toroidal magnetic field, $(|II'(\psi)| \gg |\Delta^*\psi|)$, and the toroidal current merely serves the purpose of centering the plasma. This configuration with high β_{pol}, which is sometimes referred to as a high beta tokamak or pinch, will be discussed in detail in chapter 8. The toroidal field profile as a function of the major radius for each of the three cases is illustrated in fig. 4.5.

Question 4.4.1
Can a toroidal equilibrium be confined entirely by poloidal currents? Can it be confined without externally applied fields or without image currents in the walls?

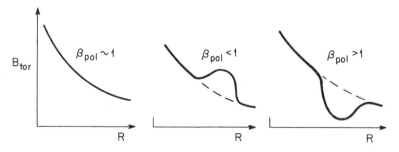

Fig. 4.5
Extreme examples of toroidal magnetic field as a function of major radius for different β_{pol}.

Table 4.1.
Reduced Equilibrium Equations

Straight Cylinder	Axisymmetric Toroid	Helical Symmetry
(x, y, z)	(R, y, φ)	(r, θ, z)
$\partial/\partial z = 0$	$\partial/\partial \varphi = 0$	$u \equiv \theta - kz$
$B = \dfrac{\partial \psi}{\partial y}\hat{x} - \dfrac{\partial \psi}{\partial x}\hat{y} + B_z \hat{z}$	$B = \dfrac{1}{R}\dfrac{\partial \psi}{\partial y}\hat{R} - \dfrac{1}{R}\dfrac{\partial \psi}{\partial R}\hat{y} + B_\varphi \hat{\varphi}$	$B_r = -\dfrac{1}{r}\dfrac{\partial \psi}{\partial u}$
		$B_\theta - kr B_z = +\dfrac{\partial \psi}{\partial r}$
$p = p(\psi)$	$p = p(\psi)$	$p = p(\psi)$
$B_z = B_z(\psi)$	$RB_\varphi = I(\psi)$	$kr B_\theta + B_z = H(\psi)$
$-\nabla^2 \psi = \mu p'(\psi) + B_z B_z'(\psi)$	$-\Delta^* \psi = R^2 \mu p'(\psi) + II'(\psi)$	$\dfrac{1}{r}\dfrac{\partial}{\partial r}\dfrac{r}{1+k^2 r^2}\dfrac{\partial \psi}{\partial r} + \dfrac{1}{r^2}\dfrac{\partial^2 \psi}{\partial u^2}$
	$\Delta^* \psi \equiv R \dfrac{\partial}{\partial R}\dfrac{1}{R}\dfrac{\partial \psi}{\partial R} + \dfrac{\partial^2 \psi}{\partial y^2}$	$= \dfrac{2kH + HH'(\psi)}{1+k^2 r^2} + \mu p'(\psi)$

MHD Equilibrium

Question 4.4.2
Can we specify any function $\psi = \psi(R, y)$ with a simple maximum and solve the Grad-Shafranov equation for the functions $p(\psi)$ and $I(\psi)$?

Question 4.4.3
A simple but exact toroidal equilibrium was suggested by Shafranov (1966) and has been extensively used by Solov'ev (1976) and others:

$$\psi = \psi_0 + \psi_1\{R^2 - a^2 y^2 - \alpha(1 - R^2)^2\} \tag{4.4.15}$$

where a^2, α, ψ_0, and ψ_1 are constants. This equilibrium follows from the Grad-Shafranov equation using the source functions

$$II'(\psi) = 2a^2 \psi_1$$
$$\mu p'(\psi) = 2(4\alpha - 1)\psi_1.$$

Any flux surface can be chosen to be the edge of the plasma (where $p = 0$), and the shape of this boundary can be adjusted by altering the parameters. Where is the magnetic axis relative to the boundary? Is the poloidal magnetic field stronger on the outer edge or the inner edge of the toroid? How elongated can the cross section be?

4.5 CYLINDER WITH ELONGATED CROSS SECTION—AN EXAMPLE OF BIFURCATION

A straight circular cylinder is one of the few equilibria which can maintain its own shape entirely by currents within the plasma. In order to deliberately change its shape, externally applied magnetic fields must be used to push or pull on the plasma. This is true if we want to elongate the plasma cross section, or bend it into a torus, or put bumps in it. A simple example will be developed in this section in order to illustrate how an externally applied field can be used to shape the plasma. It will also be shown that the shape is not uniquely determined by the external fields.

Essentially, there are two ways to elongate a current-carrying plasma—by pulling at it or by pushing on it. To pull at the plasma, an external current running in the same direction as the plasma current is used as shown in fig. 4.6a. This reduces the magnetic field, and therefore the magnetic pressure and tension on those sides of the plasma that face the external currents. The plasma responds by bulging out until the curvature and corresponding magnetic tension increase sufficiently while the pressure gradient in that direction decreases so that the forces are brought into equilibrium. The magnetic fields opposing each other

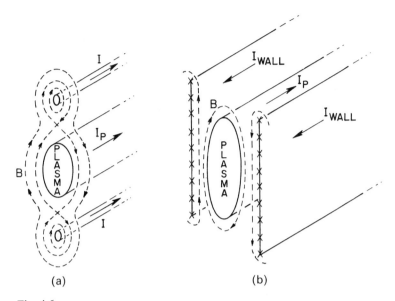

Fig. 4.6
Elongation of the plasma cross section by (a) pulling on the ends or (b) pushing on the sides.

between the plasma and the wires produce stagnation points where the poloidal magnetic field is zero. These stagnation points move closer to the plasma as the current in the wires is increased and the plasma cross section becomes more elongated. The magnetic surface running through this stagnation point, called the separatrix, usually separates the region where the plasma is well confined from the region where field lines run into the wall so that the plasma is poorly confined. Furthermore, if the external currents are made too strong, the plasma becomes unstable by shifting toward one wire or the other, given a small perturbation.

The other alternative is to push on the plasma by running a wall of current opposite to the direction of the plasma current along each side of the plasma, as in fig. 4.6b. This increases the strength of the magnetic field along the sides of the plasma so that the increased magnetic pressure squeezes the plasma. This method will be discussed in more detail in section 4.6.

Question 4.5.1
If all the external wires just discussed could be put on a circle around the plasma, a configuration pushing on the sides of the plasma could be converted into an equivalent configuration pulling at the ends

MHD Equilibrium

simply by adding a net longitudinal current uniformly distributed around the circle which, by itself, has no effect on the plasma. In view of this, is there really any difference between pushing on the plasma or pulling on it?

The simplest configuration will be used for the example to be developed here. The plasma will have the approximate form of an ellipse with uniform longitudinal current density. An externally applied quadrupole field will provide the shaping forces. This simple example can be worked out analytically, but the methods to be illustrated can be applied to more general configurations.

If a straight cylindrical plasma carries a uniform longitudinal current density, the equation for the poloidal stream function (see table 4.1) may be written

$$-\nabla^2 \psi = \mu p'(\psi) + B_z B_z'(\psi) = \mu J_{z0}. \tag{4.5.1}$$

A valid solution of this equation for the stream function and for the corresponding magnetic field is

$$\psi = \psi_0 - \frac{\mu}{2} \frac{a^2 b^2}{b^2 + a^2} J_{z0} \left(\frac{x^2}{a^2} + \frac{y^2}{b^2} \right) \tag{4.5.2}$$

$$\mathbf{B} = -\frac{a^2 b^2 \mu}{b^2 + a^2} J_{z0} \left[\frac{y}{b^2} \hat{\mathbf{x}} - \frac{x}{a^2} \hat{\mathbf{y}} \right] + B_{z0} \hat{\mathbf{z}}. \tag{4.5.3}$$

A uniform current density flows within the plasma region bounded by a flux surface, which in this case is an ellipse with major axis b and minor axis a. The pressure, or the combination of plasma pressure and longitudinal magnetic field pressure, may be specified by

$$\left. \begin{array}{l} p(\psi) \\ \\ \text{or} \\ \\ p(\psi) - \dfrac{1}{2\mu} \left(B_{z\,\text{edge}}^2 - B_z^2(\psi) \right) \end{array} \right\} = \begin{cases} J_{z0}(\psi - \psi_{\text{edge}}) \\ 0 \quad \psi < \psi_{\text{edge}} \end{cases}$$

$$= \begin{cases} \dfrac{1}{2} \dfrac{a^2 b^2}{b^2 + a^2} J_{z0} \left[1 - \left(\dfrac{x^2}{a^2} + \dfrac{y^2}{b^2} \right) \right] \\ 0 \quad \dfrac{x^2}{a^2} + \dfrac{y^2}{b^2} > 1. \end{cases} \tag{4.5.4}$$

The q-value over length $L = 2\pi/k$ can be determined by (4.3.2)

$$q(\psi) = \frac{k B_z(\psi)}{\mu J_{z0}} \frac{b^2 + a^2}{ba}. \tag{4.5.5}$$

Many researchers would be satisfied with this formal solution of a plasma equilibrium. It can be used just as it is to study the stability of the configuration and other properties of the plasma evolution such as diffusion and heat transport. However, we are concerned here with how the shape of the plasma is produced by external currents. For this purpose it is necessary to determine what part of the magnetic field is produced by currents within the plasma and what part is produced by external currents.

The field produced by the plasma currents alone can be determined by integrating the Green function over the current distribution. The Green function is simply the stream function at a point (x, y) due to a longitudinal current filament at another point (x_1, y_1)

$$\psi = \mu \frac{J(x_1, y_1)}{4\pi} \ln \frac{(x - x_1)^2 + (y - y_1)^2}{r_0^2} \qquad (4.5.6)$$

where r_0 is any radius used to make the argument of the logarithm dimensionless. After integrating (4.5.6) over the plasma current, the Taylor series expansion of the result near the magnetic axis is

$$\psi_{\substack{\text{plasma} \\ \text{currents}}} = \psi_0 - \frac{\mu}{2} \frac{J_{z0}}{b+a}(bx^2 + ay^2) + \cdots \qquad (4.5.7)$$

and the corresponding poloidal magnetic field is

$$\mathbf{B}_{\text{pol}} = \frac{\mu J_{z0}}{b+a}(bx\hat{y} - ay\hat{x}) + \cdots. \qquad (4.5.8)$$

Note that the flux surfaces due to the plasma currents alone do not conform to the shape of the plasma. Also, the magnetic field from (4.5.8) is strongest near the tips of the plasma ellipse, while the magnetic field needed for plasma equilibrium (4.5.3) is weakest there.

The magnetic field due to the plasma currents alone must be altered by the field due to a suitable arrangement of external currents until the sum of the fields just matches the field needed for equilibrium. In terms of stream functions, this procedure may be written

$$\psi_{\text{equilibrium}} = \psi_{\substack{\text{plasma} \\ \text{currents}}} + \psi_{\substack{\text{external} \\ \text{currents}}}. \qquad (4.5.9)$$

In the particular example being considered here, it suffices to use a quadrupole field, which is the lowest harmonic of the field produced by four wires running parallel to the plasma cylinder, each carrying a current I_q in alternating directions. If these wires are placed at radius r_0 from the center of the plasma, their magnetic stream function can be approximated by

$$\psi_q = \mu \frac{I_q}{\pi}\left[\left(\frac{x}{r_0}\right)^2 - \left(\frac{y}{r_0}\right)^2 + \mathcal{O}\left(\left(\frac{x}{r_0}\right)^6, \left(\frac{y}{r_0}\right)^6\right)\right]. \tag{4.5.10}$$

In order to make the proper combination of stream functions, it can be seen by substituting (4.5.2), (4.5.7), and (4.5.10) into (4.5.9) that the currents and dimensions must satisfy the following relation

$$I_q = \frac{1}{2}\frac{(b/a)^2 - 1}{(b/a)^2 + 1}\frac{1}{(b/a + 1)^2}\frac{r_0^2}{a^2}I_{\text{plasma}} \tag{4.5.11}$$

where $I_{\text{plasma}} = \pi ab J_{z0}$ is the total plasma current. This relation comes from the balance of the lowest-order multipole terms. Higher-order terms are balanced by slightly altering the plasma shape or by adding more external windings.

It can be seen from (4.5.11) that when there is no external current, there is no elongation ($b/a = 1$). As the external current is increased, the elongation is increased ($b/a > 1$). However, beyond a certain elongation ($b/a = 2.9$ from the more exact treatment by Strauss (1974)), the external current must be *decreased* to achieve further elongation. This is because the externally applied quadrupole field increases with radius more rapidly than the plasma magnetic field increases with elongation.

The important point here is that this plasma equilibrium is not uniquely determined, even if all the currents and the width of the plasma are prescribed. When the parameters are varied, there is a point of *bifurcation* at which the equilibrium can go either way. Usually the point of bifurcation separates a stable class of equilibria from an unstable class. The standard mechanical example of bifurcation is that of a steel beam under compression. Above a critical level of compression the beam can remain in equilibrium with its straight shape, or it can snap into a new equilibrium with a bowed shape. The straight shape is unstable and the bowed shape is once again stable.

Bifurcation is an important concept in the nonlinear theory of MHD instabilities. Some instabilities bend the plasma column into a helical structure which may then be a new equilibrium. When the plasma parameters are varied, the points in parameter space where the plasma is marginally stable are points of bifurcation. In the neighborhood of such points, plasma stability can be analyzed by looking for neighboring equilibria with the same constraints as the original equilibrium. This technique breaks down when the instability leads to relaxation oscillations, turbulence, or a complete disruption of the plasma.

Question 4.5.2
In the example worked out in this section, what happens if the quadrupole current is raised beyond the maximum allowed by equilibrium? Even when the current is adjusted to the equilibrium condition, are the highly elongated equilibrium states stable to further elongation?

4.6 PLASMA SQUEEZED BETWEEN CONDUCTING WALLS

When external currents are used to pull the equilibrium into elongation, as described in the last section, serious problems develop. A separatrix moves in close to the edge of the plasma, and a vertical instability develops when anything more than a mild elongation is attempted. For years researchers working with variations on this approach for simple elongations (putting aside the Doublet configuration for the moment) continued to encounter the same problems. Then, Kadish (1973) and Becker (1974) found highly elongated analytic equilibria with no separatrix at all, with what appeared to be good vertical stability, and Von Hagenow (1973) found computed equilibria with the same properties. The curious thing about Becker's first example is that he found the result fully worked out in Morse and Feshbach (1953) in the section on the "variable condenser," pp. 1247–1250. The example is quite simple, but before going into it we must understand some properties of the surface-current model for MHD equilibria.

In the *surface-current model*, all plasma currents are on the surface of the plasma. The plasma pressure is uniform within the plasma and discontinuous at the edge. The magnetic field is a vacuum magnetic field everywhere except for a discontinuity at the edge of the plasma. The discontinuities can be related to each other by considering the integral form of the equilibrium equations (4.1.8), integrated over the surface of a small pillbox straddling the edge of the plasma. Since the magnetic field is parallel to the surface of the plasma, it is parallel to the top and bottom surfaces of the arbitrarily thin pillbox, so that $\hat{n} \cdot \mathbf{B} = 0$. Then, for an arbitrarily small pillbox, (4.1.7) implies

$$(p + B^2/2\mu)_{\text{inside}} = (p + B^2/2\mu)_{\text{outside}} \tag{4.6.1}$$

where p is usually taken to be zero outside the plasma. The current density on the plasma surface $\mathbf{J}_s$ can be easily evaluated by taking a line integral of $\mathbf{J} = \nabla \times \mathbf{B}/\mu$ around a small segment of the plasma edge to obtain

$$\mu \mathbf{J}_s = \hat{n} \times (\mathbf{B}_{\text{outside}} - \mathbf{B}_{\text{inside}}) \tag{4.6.2}$$

MHD Equilibrium 77

Question 4.6.1
A physical interpretation of (4.6.1) is that the difference in plasma pressure is balanced by the difference in magnetic pressure across the plasma boundary. Why doesn't the magnetic curvature force play a role?

Question 4.6.2
Consider a circular torus in which the plasma is confined by surface currents alone. There is no poloidal field within the plasma, and the toroidal field varies like $1/R$ with a jump in magnitude across the plasma surface. All the fields can be determined analytically. How large can we make the jump in B_φ^2 before the poloidal magnetic field goes to zero at the inner edge of the torus? How does this condition scale with aspect ratio? What is the maximum $\beta \equiv P/B_{\varphi 0}^2/2\mu$ and β_{pol} that can be confined under these conditions?

Now consider the analogy between the elongated plasma equilibrium and the variable condenser problem worked out by Morse and Feshbach. Replace the potential function with the poloidal flux function and replace the electric field perpendicular to the equipotential surfaces with a magnetic field running parallel to the flux surfaces. Electric charge is replaced with longitudinal current density running into the paper. The configuration consists of a semi-infinite plate (flux surface) between two infinite plates (flux surfaces). Morse and Feshbach use conformal mapping to find the equipotential surfaces (flux surfaces) and the electric field (magnetic field) in the space between the plates. It is shown that the magnetic field is uniform along the flux surface which lies exactly halfway between the inner plate and the outer plates. Outside this flux surface nothing is changed if we take the current off the inner plate and spread it uniformly over this particular flux surface. Replacing everything within this flux surface with a uniform plasma pressure and longitudinal field, we are left with a bona fide MHD equilibrium. The plasma is semi-infinite in length, and there are no separatrixes between it and the walls. Current flows in those portions of the walls that directly face the plasma. Note that the plasma width must be half the distance between the walls.

Question 4.6.3 (A. Kadish (1973))
For a surface-current equilibrium in the form of a straight circular cylinder, how does the flux between the plasma and the wall vary as we vary the radius of the plasma to the radius of the wall, holding the B-field at the plasma fixed? Now elongate the plasma and the wall

MHD Equilibrium

holding the plasma pressure, the magnetic field at the plasma surface, and the flux between the plasma and the wall fixed. In the highly elongated form, the magnetic field between the side of the plasma and the wall is essentially uniform. What is the minimum possible width of the plasma? Slightly different widths are obtained when different shapes are used for the wall.

The computed equilibria of Von Hagenow (1973) were obtained by solving the Grad-Shafranov equation in a rectangular box with the flux ψ specified on the walls of the box (ψ = const. for the cases to be considered here). The height of the plasma is prescribed and the source functions $p'(\psi)$ and $II'(\psi)$ are set equal to zero on all flux surfaces outside the plasma. The width of the plasma, determined by the computation as a function of the height, is shown in fig. 4.7 for different current profiles within the plasma. As the plasma height shrinks to zero, the width also shrinks to zero, and the plasma has an almost circular cross section, as in the lower left of fig. 4.7. At the other extreme, the plasma touches the wall all around (the wall is a flux surface) so that the plasma is as elongated as the wall, as at the upper right in fig. 4.7. Between these extremes there is a broad plateau where the width of the plasma is essentially independent of the height. This is already evident in the 4-to-1 box elongation case shown in fig. 4.7. The plateau becomes wider as the box is further elongated. The width of the plasma in this plateau region depends upon the current profile within the plasma. If all the current is on the surface of the plasma, the plasma width is exactly half of the box width. If the current is uniform within the plasma, the plasma width is 0.73 of the box width. If the current is concentrated near the magnetic axis at the center of

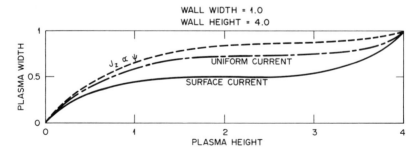

Fig. 4.7
Von Hagenow S curve showing the width of the plasma as a function of its height in a highly elongated flux box.

MHD Equilibrium

the plasma, the plasma width is essentially equal to the box width, independent of the plasma height or the box height. This means that the coils must be placed at the edge of the plasma in order to elongate it. The fact that the coils must be placed so close makes this method of elongation less attractive for controlled thermonuclear fusion experiments.

The situation is even worse when we look at the flux surfaces deep inside the plasma. As the current becomes more peaked at the magnetic axis, the central flux surfaces collapse down to circles regardless of how close the walls are placed to the plasma. This is consistent with the picture presented in fig. 4.7. The width of these inner flux surfaces is too narrow for the external currents to force them to elongate; they are in the extreme lower left of the "Von Hagenow S curves" in fig. 4.7.

4.7 TOKAMAK EQUILIBRIUM

If left to itself, a toroidal plasma will tend to expand along its *major* radius as the result of forces described and estimated later in this section. In tokamak experiments these forces are balanced by applying a vertical magnetic field to the plasma so that the cross product of the toroidal plasma current and the applied vertical field produces an inward force. If a highly conducting shell surrounds the plasma, the plasma will lean against the shell, and the resulting image current will automatically produce the required vertical field. If the vertical field is produced by a set of discrete coils (poloidal field coils), the shape of this applied field determines whether the plasma will be stable to axisymmetric vertical or radial displacements. The required vertical field and its shape will be estimated here using only space-averaged properties of the plasma. Equilibrium force balance along the *minor* radius of the plasma will not be considered here, since this depends upon the details of the pressure and current profiles as well as the shape of the externally applied fields, described in section 4.5. Of course, complete toroidal equilibria can be easily obtained by computational solution of the Grad-Shafranov equation.

The forces along the major radius are produced by (a) the plasma pressure gradient, (b) poloidal current crossed with the toroidal magnetic field, and (c) toroidal current crossed with the poloidal magnetic field. These will be considered here one at a time.

The force along the major radius due to the plasma pressure can be visualized by considering a differentially thin wedge of the torus shown in fig. 4.8. The pressures against the opposite faces of the wedge,

MHD Equilibrium

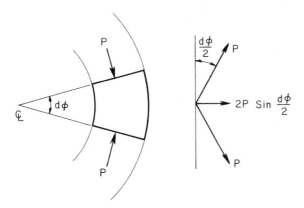

Fig. 4.8
Plasma pressure produces an outward radial force in a tokamak plasma.

integrated over the cross-sectional area, produce a net outward force

$$d\mathbf{F}_p = 2\sin(d\varphi/2) \int d\mathbf{S} \cdot \hat{\boldsymbol{\varphi}} \, p\hat{\mathbf{R}} \simeq d\varphi \int d\mathbf{S} \cdot \hat{\boldsymbol{\varphi}} \, p\hat{\mathbf{R}}.$$

The magnitude of this outward force distributed around the whole torus is

$$F_p = 2\pi \int d\mathbf{S} \cdot \hat{\boldsymbol{\varphi}} \, p(\psi). \tag{4.7.1}$$

From a particle point of view, this is a centrifugal force produced by the toroidal component of the thermal velocities.

The second radial force results from the cross product of the poloidal current in the plasma (when present) with the toroidal magnetic field. Since the toroidal magnetic field is stronger on the inner side of the torus than on the outer side ($B_\varphi = I(\psi)/R$), and any poloidal current that flows around the outer side must also flow around the inner side ($\nabla \cdot \mathbf{J} = 0$), there is a net force along the major radius from $\mathbf{J}_{pol} \times \mathbf{B}_{tor}$. The direction of this force depends upon the direction of the poloidal current; it is an outward force for a diamagnetic plasma and an inward force for a paramagnetic plasma.

The magnitude of this force can be calculated in essentially the same way that the force due to plasma pressure was calculated. Use the integral form of the equilibrium equations (4.1.8), integrated over the surface area of the very same differentially thin wedge of the torus. Each term in (4.1.8) represents the force on a surface. Consider the effect of only the toroidal magnetic field. The surface integral in (4.1.8) can be divided into a part over the faces of the wedge $d\mathbf{S} \cdot \hat{\boldsymbol{\varphi}}$ and a part over that segment of the toroidal surface that forms the edge of the plasma wedge $d\mathbf{S} \cdot \hat{\mathbf{n}}$, $\hat{\mathbf{n}} = -\nabla\psi/|\nabla\psi|$, where the toroidal field is a vacuum magnetic field $B_{\varphi\mathrm{vac}}$

$$\mathbf{F}_{B_\varphi} = -\int d\mathbf{S} \cdot \hat{\boldsymbol{\varphi}} \frac{B_\varphi^2}{2\mu} \hat{\boldsymbol{\varphi}} + \int d\mathbf{S} \cdot \hat{\mathbf{n}} \frac{B_\varphi^2}{2\mu} \hat{\mathbf{n}}. \qquad (4.7.2)$$

If there were a vacuum toroidal magnetic field throughout the plasma, there would be no net force on the plasma, and the force represented by (4.7.2) would be zero (this can be shown to be a mathematical identity (Shafranov (1971)). It follows that the forces on the faces of the wedge are due to the difference between the toroidal field within the plasma and the vacuum toroidal field

$$\mathbf{F}_{B_\varphi} = \int d\mathbf{S} \cdot \hat{\boldsymbol{\varphi}} \frac{1}{2\mu} (B_{\varphi\mathrm{vac}}^2 - B_\varphi^2) \hat{\boldsymbol{\varphi}}. \qquad (4.7.3)$$

As in the case with the force due to the plasma pressure, the net outward force around the torus due to the toroidal field alone is

$$F_{B_\varphi} = 2\pi \int d\mathbf{S} \cdot \hat{\boldsymbol{\varphi}} \frac{1}{2\mu} (B_{\varphi\mathrm{vac}}^2 - B_\varphi^2). \qquad (4.7.4)$$

Finally, an expression is needed for the force due to the toroidal current crossed with the poloidal magnetic field. An easy way to visualize this force is to imagine that the plasma is replaced by a rigid superconducting torus carrying toroidal current. Consider the magnetic field due to the plasma currents alone. The same field lines that fill all of the space on the outer side of the torus must also squeeze through the hole in the torus. Hence, the strength of this poloidal field must be larger on the inner side of the torus than on the outer side, shown in fig. 4.9, and thus there is a net outward force. By adding a vertical magnetic field, the net poloidal field strength is reduced on the inner edge of the torus and increased on the outer edge, which eliminates this force and balances the others.

The easiest way to estimate this $\mathbf{J}_{\mathrm{tor}} \times \mathbf{B}_{\mathrm{pol}}$ force is to make a virtual radial displacement of the plasma and then determine the change in the energy stored in the poloidal magnetic field due to the change in inductance, holding the poloidal flux fixed

$$\psi_{\mathrm{pol}} = LI_{\mathrm{tor}} = \text{fixed} \qquad (4.7.5)$$

$$F_{B_{\mathrm{pol}}} = -\frac{d}{dR} \tfrac{1}{2} LI_{\mathrm{tor}}^2 = \tfrac{1}{2} I_{\mathrm{tor}}^2 \frac{dL}{dR}. \qquad (4.7.6)$$

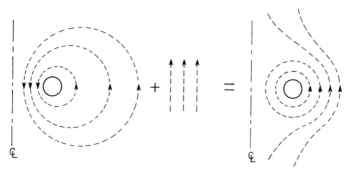

Fig. 4.9
The poloidal magnetic field from plasma current alone contributes to the outward radial force. An externally applied vertical magnetic field provides an inward force.

Question 4.7.1
What would be the force on the plasma if the toroidal current were held fixed rather than the poloidal flux? How much work would have to be done by the transformer or the poloidal field coils to hold the toroidal current fixed? Should the flux due to the applied vertical field be included in (4.7.5)?

In order to estimate this force, we can use the standard expression for the inductance of a circular torus in the large aspect ratio limit

$$L = \mu R \left(\ln \frac{8R}{a} - 2 + l_i/2 \right) \tag{4.7.7}$$

where R is the major radius, a is the minor radius, and

$$l_i = \int d\mathbf{S} \cdot \hat{\phi} B_{\text{pol}}^2 / \pi a^2 B_{\text{pol}}^2(a) \tag{4.7.8}$$

is the internal inductance per unit length of the torus—that is, the poloidal magnetic energy stored within the plasma per unit of toroidal current. For uniform and mildly peaked toroidal current distributions $l_i \geq \frac{1}{2}$.

The derivation of (4.7.7) is not straightforward. In the usual derivation the poloidal flux in the vacuum surrounding the torus is expressed as a series of special functions, and then a large aspect ratio approximation is made for the flux at the surface of the torus for use in (4.7.5). There appears to be no way to obtain the result using local expansions alone. The reason is that the total amount of poloidal flux depends upon the transition between the poloidal field near the toroidal surface,

MHD Equilibrium

where the plasma looks something like a straight wire, and the field far from the torus, where the plasma looks like a dipole source. This transition occurs at a distance comparable to the major radius of the torus—which accounts for the fact that the external inductance is nearly the same as that of a straight wire with a coaxial casing at $r = R$

$$L_{\text{ext}} = \mu R \left(\ln \frac{8R}{a} - 2 \right) \simeq \mu R \ln R/a.$$

Alternatively, special functions must be used because it is hard to find a local approximation that matches the boundary condition $\psi_{\text{pol}} = 0$ at both the center line of the torus and infinity. The large aspect ratio approximation for the inductance given by (4.7.7) is remarkably close to the exact result calculated by Malmberg and Rosenbluth (1965) and V. A. Fock (1932).

Now collect the three forces acting on the plasma along the major radius and balance them with an inward force from an applied vertical field B_y crossed with the toroidal current integrated around the torus

$$2\pi R B_y I_{\text{tor}} = 2\pi \int d\mathbf{S} \cdot \hat{\varphi} \left[p + \frac{1}{2\mu}(B^2_{\varphi\text{vac}} - B^2_\varphi) \right] + \tfrac{1}{2} I^2_{\text{tor}} \frac{dL}{dR}. \quad (4.7.9)$$

This formula is used to determine the vertical field needed to hold the plasma at any given radius. For a circular torus, the required vertical field is

$$B_y = \frac{1}{RI_{\text{tor}}} \int d\mathbf{S} \cdot \hat{\varphi} \left[p + \frac{1}{2\mu}(B^2_{\varphi\text{vac}} - B^2_\varphi) \right]$$
$$+ \frac{\mu I_{\text{tor}}}{4\pi R} \left[\ln \frac{8R}{a} - 1 + l_i/2 \right]. \quad (4.7.10)$$

Using the standard working definition of a quantity β_J similar to the poloidal beta

$$\beta_J \equiv \langle p \rangle \bigg/ \frac{1}{2\mu} \left(\oint d\mathbf{l} \cdot \mathbf{B}_{\text{pol}} \bigg/ \oint dl \right)^2 \quad (4.7.11)$$

and recalling the definition (4.4.11) of the diamagnetism μ_J, the expression for the required vertical field takes the form

$$B_y = \frac{\mu I_{\text{tor}}}{4\pi R} \left[\ln \frac{8R}{a} - 1 + \tfrac{1}{2}(l_i + \beta_J + \mu_J) \right]. \quad (4.7.12)$$

MHD Equilibrium

Question 4.7.2
Consider a high-beta tokamak in which $\beta_J \simeq \mu_J \gg 1$. For a circular torus and uniform vertical field, how large can β_J be made before the vertical field cancels the poloidal field at the inner edge of the torus? What is the corresponding limit on

$$\langle \beta \rangle \equiv \langle p \rangle \bigg/ \frac{1}{2\mu} B_{\varphi 0}^2 ? \tag{4.7.13}$$

What happens when we try to make β_J larger?

Question 4.7.3 (Clarke and Sigmar (1977))
Suppose a perfectly conducting torus were heated so that β was raised while $q(\psi)$ remained fixed. Could the vertical field cancel the poloidal field now that the q-value is held fixed at the surface of the plasma? Under these conditions, can β be raised indefinitely? What would the poloidal fields look like to accomplish this?

Question 4.7.4
The TO-1 tokamak in the Soviet Union has been operated with the vertical field activated around only one quarter of the torus, with no major change in the plasma behavior. How are the forces resolved? How large must this vertical field be relative to the field activated around the whole torus under the same operating conditions?

While we are on the subject of the vertical field needed for equilibrium, it would be useful to determine the shape this field should have in order to avoid simple axisymmetric instabilities. If the vertical field were uniform, straight up and down, the equilibrium would be neutrally stable—it could be shifted either up or down without producing a restoring force.

If the vertical field curves inward toward the center line of the torus, shown in fig. 4.10, the plasma is stabilized with respect to rigid vertical displacements. To see this, perturb the plasma upward or downward and observe that the toroidal plasma current crossed with that small component of the vertical field that points toward the center line of the torus produces a vertical restoring force. If the vertical field curves away from the center line, the plasma will be unstable with respect to these vertical displacements. This curvature of the externally applied vacuum magnetic field is most conveniently described by the *decay index* of the field defined by

MHD Equilibrium

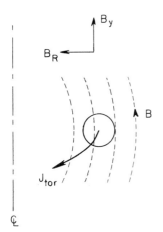

Fig. 4.10
Shape of the vertical field needed for axisymmetric stability.

$$n \equiv -\frac{R}{B_{vert}} \frac{d}{dR} B_{vert}. \tag{4.7.14}$$

The condition for vertical stability is

$$0 < n. \tag{4.7.15}$$

For horizontal stability we must require that the restoring force from the vertical field decrease slower with radius than the expansion forces from the plasma

$$\frac{d}{dR}(2\pi R I B_y) > \frac{d}{dR}\left\{\tfrac{1}{2}\mu I^2 \left[\ln\frac{8R}{a} - 1 + \tfrac{1}{2}(l_i + \beta_J + \mu_J)\right]\right\}.$$

As before, the poloidal flux must be held fixed under the radial derivative—but now the flux from the vertical field must be included

$$\psi = LI - 2\pi \int_0^R dR\, R B_y. \tag{4.7.16}$$

In the limit of large aspect ratio, $\ln(8R/a) \gg 1$, the condition on the field index required for horizontal stability turns out to be

$$n < 3/2. \tag{4.7.17}$$

Question 4.7.5
Note that for the horizontal stability result (4.7.17) we did not require that the toroidal flux or the entropy be conserved in the virtual dis-

placement. What would the result be if these constraints were implemented?

Question 4.7.6 (Bartoli and Green (1963))
The toroidal tokamak plasma can be regarded as a magnetic dipole aligned against the vertical magnetic field. By flipping over, the dipole can go into a state of lower energy. Why doesn't the tokamak plasma tend to flip over as a result of the torque exerted upon it by the vertical magnetic field?

It is a pleasure to acknowledge the help of Lee Berry and A. T. Mense in preparing section 4.7.

4.8 SUMMARY

The basic equilibrium force balance equations are (4.1.1) to (4.1.3). However, the Grad-Shafranov equation (4.4.10) is almost always used to compute equilibria in axisymmetric toroidal configurations, as are its counterparts for straight cylindrical and helically symmetric configurations in table 4.1. The pressure and total poloidal current in these equations must be given as functions of the stream function ψ, which is proportional to the poloidal flux, rather than as functions of space.

Externally applied magnetic fields must be used if the plasma is to be deformed from a straight circular cylindrical shape. The external current required to elongate the plasma with a quadrupole field is estimated in (4.5.11). The quadrupole currents can be arbitrarily far away, but only small elongations can be achieved. Arbitrary elongations can be achieved by squeezing the plasma between walls of current running opposite to the plasma current, but the walls must be very close to the plasma, and the current profile within the plasma must be broad.

The vertical field needed to prevent a circular toroidal plasma from expanding along the major radius is given by (4.7.10). The derivation of this vertical field in section 4.7 illustrates how integral equilibrium properties are used.

4.9 REFERENCES

For the mathematical groundwork of MHD equilibrium theory, see
M. D. Kruskal and R. M. Kulsrud, *Physics of Fluids*, *1*, 265–274 (1958).

An additional section was added to this paper in *Plasma Physics*

(Seminar at Trieste), IAEA, Vienna, 1965.

H. Grad and H. Rubin, IAEA Geneva Conf., *31*, 190–197 (1958).

V. D. Shafranov has probably published more than anyone else on MHD equilibrium theory. He has three review articles:

V. D. Shafranov, *Reviews of Plasma Physics*, 2, 103–151 (New York: Consultants Bureau, 1966).

L. S. Solov'ev and V. D. Shafranov, *Reviews of Plasma Physics*, *5*, 1–248 (1970).

V. S. Mukhovatov and V. D. Shafranov, *Nuclear Fusion*, *11*, 605–633 (1971).

Some of his more recent publications are

V. D. Shafranov, *Plasma Physics*, *13*, 757–762 (1971).

L. E. Zakharov and V. D. Shafranov, *Sov. Phys. Tech. Phys.*, *18*, 151–156 (1973).

V. D. Shafranov, Proc. IAEA Workshop on Fusion Reactor Design Problems, (Culham, 1974), pp. 249–259, IAEA Vienna, 1974.

An alternative derivation of the Grad-Shafranov equation can be found in the book by

W. B. Thompson, *An Introduction to Plasma Physics* (Oxford: Pergamon Press, 1962).

A good review of computational methods used to solve the Grad-Shafranov equation can be found in the article by

R. W. Hockney, *Methods of Computational Physics*, *9*, 135–211 (New York: Academic Press, 1970).

Recently, an infinite number of points of bifurcation were found for the equilibrium treated in section 4.5 by

J. C. B. Pabaloizou, I. Rebelo, J. J. Field, C. L. Thomas, and F. A. Haas, *Nuclear Fusion*, *17*, 33–46 (1977).

The same equilibrium has also been treated by

H. R. Strauss, *Phys. Fluids*, *17*, 1040–1041 (1974).

R. Gajewski, *Phys. Fluids*, *15*, 70–74 (1972).

Some references for section 4.6 are

G. Becker, *Nuclear Fusion*, *14*, 319–321 (1974).

K. U. Von Hagenow and K. Lackner, Proc. 3rd Int. Symposium on Toroidal Plasma Confinement, Garching, 1973, F-7.

A. Kadish, *Nuclear Fusion*, *13*, 756–757 (1973).

A. Kadish and D. C. Stevens, *Nuclear Fusion*, *14*, 821–829 (1974).

W. Feneberg and K. Lackner, *Nuclear Fusion*, *13*, 549–556 (1973).

The simplified derivation used in section 4.7 for the vertical field needed for a toroidal equilibrium came from

R. G. Mills, Princeton Plasma Phys. Lab. report MATT-800 (Nov. 1970).

A. T. Mense, Univ. of Wisconsin, Nuclear Engr. Dept. report FDM-71 (Sept. 1973).

More sophisticated analysis can be found in the review articles cited above and in

J. M. Greene, J. L. Johnson, K. E. Weimer, *Phys. Fluids, 14*, 671–683 (1971).

S. Yoshikawa, *Phys. Fluids, 7*, 278–283 (1964).

The inductance for a circular torus is computed by

J. H. Malmberg and M. N. Rosenbluth, *Rev. Sci. Inst., 36*, 1886–1887 (1965).

V. A. Fock, *Phys. Z. Sowjetunion, 1*, 215 (1932).

Some recent references on tokamaks with high poloidal beta are

J. D. Callen and R. A. Dory, *Phys. Fluids, 15*, 1523–1528 (1972).

J. F. Clarke and D. J. Sigmar, *Phys. Rev. Letters, 38*, 70–74 (1977).

R. A. Dory and Y-K. M. Peng, *Nuclear Fusion, 17*, 21–31 (1977).

Other references cited in the text are

G. Bateman, *Nuclear Fusion, 13*, 581–594 (1973).

L. S. Solov'ev, *Reviews of Plasma Physics, 6*, 239–331 (New York: Consultants Bureau, 1976).

P. M. Morse and H. Feshbach, *Methods of Theoretical Physics* (New York: McGraw-Hill, 1953).

C. Bartoli and T. S. Green, *Nuclear Fusion, 3*, 84–88 (1963).

5
Linearized Equations and the Energy Principle

Most of the MHD instability literature deals with the linearized MHD equations in order to determine whether a given equilibrium is stable or unstable to arbitrarily small perturbations. The vast majority of these authors carry out their mathematical analyses using a variational form of the equations called the *energy principle*. Although this variational technique is very powerful and often simplifies the analysis, there are subtleties that should be understood before it is used. This chapter emphasizes the interpretation and derivation of the various forms of the energy principle as well as listing the alternative mathematical methods used to study linear instabilities.

At this point it is worthwhile to review and make more precise our definitions of equilibrium and stability. MHD equilibrium is the complete balance of forces at every point in space. *Instability* is the tendency to move away from equilibrium, given an arbitrarily small velocity perturbation. *Stability* is the tendency to return to the original equilibrium. Strictly speaking, for the purposes of testing for instability, the perturbed state must be in equilibrium initially or at some instant in time. Hence, a velocity perturbation is used rather than a perturbation in pressure or magnetic field. In the analogy of a ball in the bottom of a bowl, this caveat excludes perturbations where the ball rolls around without ever passing through the bottom.

The concept of *marginal stability* is more difficult to define; roughly speaking, it is the borderline between stability and instability. A problem arises when there is a degeneracy in the equilibrium. For example, suppose a ball is set into motion around a circular trough in the bottom of a bowl. Is this state stable or only marginally stable? The distinction becomes less obvious in a fluid continuum which has an infinite number of degrees of freedom—three for each point in the fluid.

Question 5.1
Consider the following subtleties in the definition of stability:
a. Suppose the plasma is given a velocity perturbation which does not change the shape, fields, pressure, or forces within the plasma in any way. The velocity continues unchanged. Is this plasma marginally stable?
b. Suppose the perturbed plasma tends to return to the original equilibrium, but the restoring forces are so great that the plasma overshoots each time with an ever-growing perturbation (overstability). Is this plasma stable?
c. Suppose the perturbed plasma tries to return to the original equilibrium but viscosity or dissipation force it to come to rest at a new equilibrium. Is this plasma unstable?
d. Suppose the plasma breaks into convection cells which remain in steady state. Is this new state of the plasma unstable?
e. Suppose there is a source of energy which drives random turbulence which is in steady state only in the average. Is this turbulent state unstable?

5.1 LINEARIZED EQUATIONS

Our starting point will be the system of ideal MHD equations (2.1.1)–(2.1.6) together with boundary conditions (2.5.5) and (2.5.7) which conserve mass, magnetic flux, and energy in the system. These equations are linearized by considering an arbitrarily small perturbation from a stationary equilibrium specified by (4.1.1)–(4.1.3) and $\mathbf{v}^0(\mathbf{x}) = 0$, $\rho^0 = \rho^0(\mathbf{x})$. All equilibrium quantities will be denoted by a superscript 0 and all perturbed quantities by a superscript 1. The linearized MHD equations are

$$\rho^0 \frac{\partial \mathbf{v}^1}{\partial t} = -\nabla p^1 + \mathbf{J}^0 \times \mathbf{B}^1 + \mathbf{J}^1 \times \mathbf{B}^0 \qquad (5.1.1)$$

$$\mathbf{J} = \frac{1}{\mu} \nabla \times \mathbf{B}$$

Linearized Equations and the Energy Principle

$$\frac{\partial \mathbf{B}^1}{\partial t} = \nabla \times (\mathbf{v}^1 \times \mathbf{B}^0) \qquad (5.1.2)$$

$$\frac{\partial p^1}{\partial t} = -\mathbf{v}^1 \cdot \nabla p^0 - \Gamma p^0 \nabla \cdot \mathbf{v}^1, \qquad \Gamma = 5/3. \qquad (5.1.3)$$

The equation for the perturbed density

$$\frac{\partial \rho^1}{\partial t} = -\mathbf{v}^1 \cdot \nabla \rho^0 - \rho^0 \nabla \cdot \mathbf{v}^1 \qquad (5.1.4)$$

is not needed because the perturbed density does not appear in any of the other equations. Instabilities in moving equilibria ($\mathbf{v}^0 \neq 0$) will not be considered here.

Note that the magnitude of the seven perturbation variables ($\mathbf{v}^1$, $\mathbf{B}^1$, p^1) is immaterial here since the linearization may be written with an arbitrarily small scale factor ε muliplying the perturbation

$$\begin{pmatrix} \mathbf{v} \\ \mathbf{B} \\ p \\ \rho \end{pmatrix} = \begin{pmatrix} 0 \\ \mathbf{B}^0 \\ p^0 \\ \rho^0 \end{pmatrix} + \varepsilon \begin{pmatrix} \mathbf{v}^1 \\ \mathbf{B}^1 \\ p^1 \\ \rho^1 \end{pmatrix}. \qquad (5.1.5)$$

The fundamental reason for linearizing the MHD equations is that any perturbation may then be expressed as the sum of eigenfunctions, and these eigenfunctions are unique for any given equilibrium and boundary conditions—independent of the choice of initial perturbation. The initial conditions determine only how much of each eigenfunction is present for that particular perturbation. Since the coefficients of the perturbation equations (5.1.1)–(5.1.4) are constant in time, when linearizing about a static equilibrium, all of the eigenfunctions have a simple exponential time dependence—with the exception of degenerate eigenfunctions, which may have algebraic time dependence.

5.2 ξ-FORM OF THE EQUATIONS

A useful transformation can be accomplished by writing the linearized equations in terms of the *displacement vector*

$$\boldsymbol{\xi}(\mathbf{x}, t) \equiv \int_0^t dt' \, \mathbf{v}^1(\mathbf{x}, t'). \qquad (5.2.1)$$

If we were using a Lagrangian coordinate system moving with the fluid, $\boldsymbol{\xi}$ would represent the displacement of a fluid element from its initial position. But here we are using Eulerian coordinates which are fixed in an inertial frame, so that the evolution of ξ characterizes the flow of a succession of different fluid elements past a fixed point. The derivation of the fixed-boundary energy principle is simpler in Eulerian coordinates while free boundaries are probably handled better with Lagrangian coordinates. The vector field ξ is well-defined in either coordinate system and the distinction in interpretation is not relevant as long as the perturbation is arbitrarily small.

The linearized equations (5.1.2) and (5.1.3) for the perturbed magnetic field and pressure can be integrated once to yield

$$\mathbf{B}^1(\mathbf{x}, t) = \nabla \times (\xi \times \mathbf{B}^0) + \mathbf{B}^1(\mathbf{x}, 0) \tag{5.2.2}$$

$$p^1(\mathbf{x}, t) = -\xi \cdot \nabla p^0 - \Gamma p^0 \nabla \cdot \xi + p^1(\mathbf{x}, 0). \tag{5.2.3}$$

The constants of integration in (5.2.2) and (5.2.3) may be eliminated *if the perturbation represents a dynamic state which has passed through the equilibrium.* This is a standard assumption in the literature. Dynamically, this assumption corresponds to starting with a ball at the top of a hill with some initial velocity rather than starting with the ball displaced from the top of the hill along some trajectory that may never pass over the top. For the linearized MHD equations, this assumption is implemented by using the initial conditions

$$\xi(\mathbf{x}, 0) = 0 \qquad \frac{\partial}{\partial t}\xi(\mathbf{x}, 0) \neq 0 \tag{5.2.4}$$

$$\mathbf{B}^1(\mathbf{x}, 0) = 0 \qquad p^1(\mathbf{x}, 0) = 0.$$

The perturbed magnetic field and pressure then become

$$\mathbf{B}^1(\mathbf{x}, t) = \nabla \times (\xi \times \mathbf{B}^0) \tag{5.2.5}$$

$$p^1(\mathbf{x}, t) = -\xi \cdot \nabla p^0 - \Gamma p^0 \nabla \cdot \xi. \tag{5.2.6}$$

The linearized equations may then be combined into a single second-order partial differential equation for the displacement vector

$$\rho^0 \frac{\partial^2 \xi}{\partial t^2} = \mathbf{F}\{\xi\} \tag{5.2.7}$$

$$= \nabla(\xi \cdot \nabla p^0 + \Gamma p^0 \nabla \cdot \xi)$$

$$+ \frac{1}{\mu}(\nabla \times \mathbf{B}^0) \times [\nabla \times (\xi \times \mathbf{B}^0)]$$

$$+ \frac{1}{\mu}(\nabla \times [\nabla \times (\xi \times \mathbf{B}^0)]) \times \mathbf{B}^0.$$

Linearized Equations and the Energy Principle

This system may be isolated from the rest of the world by using the boundary condition

$$\xi_\perp = 0 \qquad (5.2.8)$$

along an equilibrium magnetic surface at the edge of the plasma or by allowing for a vacuum region bounded by a perfectly conducting wall. Equation (5.2.7) is usually the starting point for linear stability analyses.

Question 5.2.1
The original linearized MHD equations (5.1.1)–(5.1.3) were a seventh-order system of equations for $\mathbf{v}^1$, $\mathbf{B}^1$ and p^1. Equation (5.2.7) is a sixth-order system of partial differential equations for ξ. Where did the seventh order go?

5.3 THE ENERGY PRINCIPLE

The energy principle can be derived in much the same way as the variational forms were derived in the chapter on the Rayleigh-Taylor instability. Multiply the equation of motion (5.2.7) by the time derivative of the displacement vector $\dot{\xi}$ and integrate over the volume of the plasma with boundary condition (5.2.8)

$$\int d^3x \rho^0 \ddot{\xi} \cdot \dot{\xi} = \frac{\partial}{\partial t} \int d^3x \tfrac{1}{2}\rho \dot{\xi}^2 = \int d^3x \dot{\xi} \cdot \mathbf{F}\{\xi\}. \qquad (5.3.1)$$

The system is said to be *self-adjoint* if we can integrate by parts to demonstrate the identity

$$\int d^3x\, \boldsymbol{\eta} \cdot \mathbf{F}\{\xi\} = \int d^3x\, \xi \cdot \mathbf{F}\{\boldsymbol{\eta}\} \qquad (5.3.2)$$

for any choice of η and ξ subject to the boundary condition. This integration by parts is quite involved; it has been carried out in detail by Kadomtsev (1963, pp. 158–162) and Kulsrud (1964, pp. 67–72) among others. Using this property of self-adjointness, it follows from (5.3.1) that the total perturbed energy is constant in time

$$\frac{\partial}{\partial t}\left[\int d^3x \tfrac{1}{2}\rho^0 \dot{\xi}^2 - \tfrac{1}{2}\int d^3x\, \xi \cdot \mathbf{F}\{\xi\}\right] = 0 \qquad (5.3.3)$$

kinetic + potential energy

In the literature, the potential energy is denoted by

$$\delta W \equiv -\tfrac{1}{2}\int d^3x\, \xi \cdot \mathbf{F}\{\xi\}. \qquad (5.3.4)$$

From this constancy of the perturbed energy, it follows that any perturbation that decreases the potential energy (makes δW negative) produces a corresponding increase in the kinetic energy which indicates that the system is linearly unstable. That is, the initial velocity perturbation increases. Such a perturbation does not need to be the fastest growing instability or an eigenfunction of the equations or a state that is likely to be reached during the dynamic evolution of the system. Any test function will do, as long as it satisfies the boundary conditions and is integrable. This is one of the advantages of using (5.3.4) to test for instability.

Alternatively, if all perturbations lead to an increase in the potential energy (δW positive), then the system is linearly stable to exponentially growing modes. A rigorous proof of this statement is contained in papers by Spies (1974) and Laval, Mercier, and Pellat (1965).

Starting with the energy principle, the linear equations of motion (5.2.7) can be derived using the principle of least action

$$\delta \int dt\, L \equiv \delta \int dt\, [\text{K.E.} - \delta W(\xi, \xi)] = 0 \tag{5.3.5}$$

where the variational $\delta \xi(\mathbf{x}, t)$ is arbitrary except at the endpoints of the time integration. Hence, the equation of motion is the Euler-Lagrange equation for the energy principle.

Question 5.3.1
Starting with the conservation of energy for the nonlinear MHD equations

$$E = \int d^3x \left(\tfrac{1}{2}\rho v^2 + \frac{p}{\Gamma - 1} + \frac{1}{2\mu} B^2 \right), \tag{5.3.6}$$

which is the integral form of (2.5.4) for the boundary conditions being used here, can you derive the conservation of energy (5.3.3) for the linearized equations? What about the second-order pressure and magnetic field?

Since the linearized equations of motion (5.2.5) have constant coefficients in time, they can be written as an eigenvalue equation, as in chapter 3,

$$\xi(\mathbf{x}, t) = \text{Re}\{\xi(\mathbf{x}) e^{\gamma t}\} \tag{5.3.7}$$

$$\gamma^2 \rho^0 \xi(\mathbf{x}) = \mathbf{F}\{\xi(\mathbf{x})\} \tag{5.3.8}$$

where each eigenvalue γ and eigenfunction $\xi(\mathbf{x})$ may be complex-valued.

Linearized Equations and the Energy Principle

(Often $i\omega$ is used in the literature rather than γ.) It follows directly from the self-adjointness of the real-valued operator F that γ^2 must be real-valued

$$(\gamma^2)^* \rho^0 \xi^* = \mathbf{F}\{\xi^*\}$$

$$\gamma^2 \int d^3x \, \rho^0 \xi^* \cdot \xi = \int d^3x \, \xi^* \cdot \mathbf{F}\{\xi\} = \int d^3x \, \xi \cdot \mathbf{F}\{\xi^*\}$$

$$= (\gamma^2)^* \int d^3x \, \rho^0 \xi \cdot \xi^*$$

hence $\gamma^2 = (\gamma^2)^*$.

This means that γ is either real-valued, corresponding to exponential growth or decay, or purely imaginary, corresponding to oscillation. There are no eigenfunctions in the ideal MHD model that correspond to growing or damped oscillations. Eigenfunctions may be combined, however, to obtain growing or damped oscillations over a limited time interval, just as the eigenfunctions of the Vlasov equation (Case-Van Kampen modes) can be combined to demonstrate Landau damping.

Question 5.3.2
The eigenfunction corresponding to a linear instability has the form $(\mathbf{v}^1, \mathbf{B}^1, p^1)(\mathbf{x}, t) = (\mathbf{v}^1, \mathbf{B}^1, p^1)(\mathbf{x})e^{\gamma t}$. How can such an eigenfunction represent a state that has passed through the equilibrium ($\mathbf{v}^1 \neq 0$, $\mathbf{B}^1 = 0$, $p^1 = 0$) at $t = 0$?

Question 5.3.3
Given any equilibrium, show that any motion of the form $\mathbf{v}^1 = \alpha(p^0)\mathbf{B}^0 + \beta(p^0)\mathbf{J}^0$ does not change the magnetic field or pressure and, therefore, does not induce any force. Does this mean that every equilibrium is at least marginally unstable? Does this perturbation represent a state that has passed through the equilibrium?

5.4 DIFFERENT FORMS OF THE ENERGY PRINCIPLE

There are many ways to rewrite the potential energy $\delta W(\xi, \xi)$: integrating by parts, breaking terms down into components, adding terms that integrate to zero for the given boundary conditions, etc. For example, an important form of the energy principle is the symmetric form in which ξ appears only in factors that are squared. The derivation of this symmetric form demonstrates the self-adjoint nature of the force operator (5.2.7). By examining different forms of the energy principle, it is possible to gain insight into the effects that tend to drive instabilities (negative terms in δW) and the effects that tend to stabilize instabilities (positive terms in δW). Here we shall list a number of different ways of writing the energy principle.

From (5.2.7) and (5.3.4), the basic form of the potential energy within the plasma is

$$\delta W = -\tfrac{1}{2}\int d^3x\,\xi\cdot \mathbf{F}\{\xi\} \qquad (5.4.1)$$
$$= -\tfrac{1}{2}\int d^3x\{\nabla[\xi\cdot\nabla p^0 + \Gamma p^0 \nabla\cdot\xi] + \mathbf{J}^0\times\mathbf{B}^1 + \mathbf{J}^1\times\mathbf{B}^0\}\cdot\xi$$

where $\mathbf{J} = \nabla\times\mathbf{B}$, $\quad \mathbf{B}^1 = \nabla\times(\xi\times\mathbf{B}^0)$, $\quad$ and $\quad \Gamma = 5/3$.

The p^1 and $\mathbf{J}^1\times\mathbf{B}^0$ terms can be integrated by parts over the plasma volume to produce

$$\delta W_F = \tfrac{1}{2}\int d^3x\left\{\frac{1}{\mu}|\mathbf{B}^1|^2 + \mathbf{J}^0\cdot\xi\times\mathbf{B}^1 + \Gamma p^0|\nabla\cdot\xi|^2 \right.$$
$$\left. + \xi\cdot\nabla p^0\nabla\cdot\xi\right\} \qquad (5.4.2)$$

plus a surface term

$$\delta W - \delta W_F = \tfrac{1}{2}\int d\mathbf{S}\cdot\xi(p^1 + \mathbf{B}^0\cdot\mathbf{B}^1). \qquad (5.4.3)$$

If the plasma extends to a rigid wall where $d\mathbf{S}\cdot\xi = 0$, this surface term is zero. If the plasma is surrounded by a vacuum, this surface term can be integrated by parts into the vacuum region to produce an extended form of the energy principle

$$\delta W = \delta W_F + \delta W_V + \delta W_S \qquad (5.4.4)$$

where δW_F is the potential energy from the perturbation of the fluid given by (5.4.2), while

$$\delta W_V = \tfrac{1}{2}\int_{\text{vacuum}} d^3x\,\frac{1}{\mu}|\mathbf{B}^1|^2 \qquad (5.4.5)$$

is the perturbation of the magnetic energy in the vacuum region, and

$$\delta W_S = \tfrac{1}{2}\int d\mathbf{S}\cdot\left[\nabla\left[p^0 + \frac{1}{2\mu}(B^0)^2\right]\right](\hat{\mathbf{n}}\cdot\xi)^2 \qquad (5.4.6)$$

is a surface integral that vanishes unless the equilibrium has surface currents. The notation $[\![A]\!]$ means the discontinuity in A across the plasma surface. This extension of the energy principle into the vacuum region was demonstrated by Bernstein, Frieman, Kruskal, and Kulsrud (1958, p. 23). Their derivation is given in more detail by Schmidt (1966, pp. 119–125).

The expression (5.4.2) for the potential energy within the plasma can be further rearranged and integrated by parts to produce the symmetric form first derived by Bernstein, Frieman, Kruskal, and Kulsrud (1958, p. 30)

$$\delta W_F = \tfrac{1}{2} \int_{\text{plasma}} d^3x \left\{ \frac{1}{\mu} |\mathbf{B}^1 + \mu \hat{\mathbf{n}} \cdot \xi \mathbf{J}^0 \times \hat{\mathbf{n}}|^2 \right. \tag{5.4.7}$$
$$\left. + \Gamma p^0 |\mathbf{V} \cdot \xi|^2 - 2\mathbf{J}^0 \times \hat{\mathbf{n}} \cdot (\mathbf{B}^0 \cdot \mathbf{V}\hat{\mathbf{n}})(\hat{\mathbf{n}} \cdot \xi)^2 \right\}$$

where $\hat{n} = -\nabla p^0/|\nabla p^0|$ is the unit vector normal to the equilibrium magnetic surfaces. Kadomtsev (1963, pp. 158–162) and Kulsrud (1964, pp. 67–72) carry out the details of this derivation. Only the third term in (5.4.7) can be negative and therefore destabilizing. Liley (1962) writes this third term as the sum of the normal curvature κ and the torsion $\tau = -\hat{\mathbf{B}} \cdot \mathbf{V}(\hat{\mathbf{B}} \times \kappa/\kappa)$ of the equilibrium magnetic field

$$\mathbf{J}^0 \times \hat{\mathbf{n}} \cdot \mathbf{V}\hat{\mathbf{n}} \cdot \mathbf{V} B^0 = |\nabla p^0|(\kappa + \tau \cot \theta) \tag{5.4.8}$$

where θ is the angle between $\mathbf{J}^0$ and $\mathbf{B}^0$.

Perhaps the most illuminating way to write the potential energy integral is

$$\delta W_F = \tfrac{1}{2} \int_{\text{plasma}} d^3x \left\{ \frac{1}{\mu} |\mathbf{B}_\perp^1|^2 + \mu \left| \frac{1}{\mu} \mathbf{B}_\parallel^1 - \mathbf{B}^0 \xi \cdot \nabla p^0/|B^0|^2 \right|^2 + \Gamma p^0 |\mathbf{V} \cdot \xi|^2 \right.$$
$$\quad\quad\quad\text{Alfvén} \quad\quad\quad\quad \text{fast magnetoacoustic} \quad\quad\quad \text{acoustic}$$
$$\left. + \frac{\mathbf{J}^0 \cdot \mathbf{B}^0}{|B^0|^2} \mathbf{B}^0 \times \xi \cdot \mathbf{B}^1 - 2\xi \cdot \nabla p^0 \, \xi \cdot \kappa \right\}. \tag{5.4.9}$$
$$\quad\quad\quad\text{kink} \quad\quad\quad\quad\quad \text{interchange}$$

This is similar to the form first written down by Furth, Killeen, Rosenbluth, and Coppi (1965) and by Greene and Johnson (1968). Each term in (5.4.9) has a physical significance. The first term of the integrand is the magnetic energy in the Alfvén wave (sometimes called the shear Alfvén wave). The second term, which can be written in the form

$$\tfrac{1}{2} \int_{\text{plasma}} d^3x \frac{(B^0)^2}{\mu} \left| \mathbf{V} \cdot \xi + \frac{\xi \cdot \mathbf{V}[(B^0)^2 + 2\mu p^0]}{(B^0)^2} \right|^2, \tag{5.4.10}$$

is the potential energy of the fast magnetoacoustic wave (or fast Alfvén wave). The $\Gamma p^0 |\nabla \cdot \xi|^2$ term is the potential energy associated with ordinary sound waves. The next term, when it is negative, can drive the so-called kink instabilities (or current-driven instabilities). The final term can drive *interchange* or *ballooning* instabilities that correspond to a Rayleigh-Taylor instability driven by pressure gradient and curvature. We will deal with the details of these instabilities in the next three chapters.

5.5 METHODS USED IN LINEAR STABILITY ANALYSIS

A variety of methods have been developed to test equilibria for stability and to study the nature of instabilities when they are present. Here we shall list some of the techniques and indicate how they are used.

1. The linearized equations of motion can be advanced in time as an initial boundary-value problem. If the equilibrium is unstable, the fastest growing instability will dominate over all other motion after several *e*-folding times for almost any initial perturbation. Computer programs using this technique were developed by Bateman, Schneider, and Grossmann (1974) and by Wesson and Sykes (1974). The results of these studies will be discussed in section 7.4.

2. Equation (5.3.8), which is actually three equations for the three components of ξ, can be solved as an eigenvalue problem for the spectrum of eigenvalues and their corresponding eigenfunctions. This method is especially usefull for the straight circular cylinder (Grossmann and Ortolani (1973)) where the equations can be reduced to a pair of ordinary differential equations along one dimension, as we shall see in section 6.3.

3. A test function with adjustable parameters can be used for each component of ξ in an effort to maximize the growth rate in the variational form of the eigenvalue equations

$$\delta[\gamma^2 K\{\xi,\xi\} + \delta W\{\xi,\xi\}] = 0 \qquad (5.5.1)$$

where

$$K\{\xi,\xi\} \equiv \int_{\text{plasma}} d^3x \, \rho^0 \xi^2 \qquad (5.5.2)$$

This procedure can be thought of as a minimization of δW using the normalization $K\{\xi,\xi\} = 1$ as a constraint to keep ξ nontrivial. In this interpretation, γ^2 is the undetermined Lagrange multiplier used to find the constrained minimum. This direct variational method was used by

Linearized Equations and the Energy Principle 99

Kerner and Tasso (1974) to study the instabilities of Shafranov's analytic toroidal equilibrium (see question 4.4.3). Also, this method is at the heart of the finite element codes used to determine the complete spectrum of MHD eigenvalues for any axisymmetric equilibrium (Grimm et al. (1976), Appert, Gruber, and Troyon (1975)).

4. If we are willing to be satisfied with less information, we can change the rules of the game to suit our own purposes or to simplify the problem. For example, if we wish merely to test for stability, without regard to growth rates or oscillation frequencies, it is sufficient to minimize the potential energy δW and ignore the contribution from the kinetic energy. The normalization $K\{\xi,\xi\} = 1$ can then be replaced with any other normalization or constraint. By doing this we lose the ability to estimate growth rates reliably or to compute the structure of eigenfunctions.

A technique frequently used is to find upper and lower bounds on γ^2 by selectively eliminating negative or positive definite terms from any of the expressions for $\delta W\{\xi,\xi\}$. In this way one obtains necessary or sufficient conditions for stability (Solov'ev (1971), Lortz (1973)).

5. When doing analysis, it is often useful to obtain a partial minimization of the potential energy. A full minimization of δW is equivalent to finding a solution of the Euler equations for the three components of ξ. In many problems treated in the literature it often happens that one or two of these Euler equations are simple enough to be solved explicitly. This happens, for example, when a specially restricted normalization is used or when there is enough symmetry in the equilibrium to make effective use of Fourier analysis. An excellent example of partial minimization will be considered in the derivation of the Mercier criterion in section 7.2.

Question 5.5.1

Suppose an equilibrium is marginally stable, and we perturb it with only the velocity field part of a marginally stable eigenfunction. How will the perturbed pressure and magnetic field behave as a function of time?

I would like to acknowledge the aid of many enlightening discussions with David Nelson during the preparation of this chapter.

5.6 SUMMARY

The linearized equations for a small perturbation around a stationary, static equilibrium are usually written in terms of the fluid displacement vector defined by (5.2.1). It is standard to assume that the fluid has an initial velocity perturbation, but the initial perturbations in the displacement, magnetic field, and pressure are all zero (5.2.4). Then the linearized equations of motion for the displacement are (5.2.7), the perturbed magnetic field and pressure given by (5.2.5) and (5.2.6).

δW denotes the potential energy associated with a small perturbation (5.3.4). If any perturbation ξ makes δW negative, there is an exponentially growing instability. If all nontrivial perturbations make δW positive, the plasma is MHD stable. All nondegenerate eigenfunctions grow exponentially or oscillate in time with constant amplitude (γ^2 is real-valued).

The expression for δW can be extended into a vacuum region (5.4.2)–(5.4.6). The two most useful forms for the fluid part of δW_F are given by (5.4.7) and (5.4.9).

5.7 REFERENCES

I. B. Bernstein, E. A. Frieman, M. D. Kruskal, and R. M. Kulsrud, *Proc. Roy. Soc.*, *A244*, 17–40 (1958); also reprinted in *MHD Stability and Thermonuclear Containment*, A. Jeffrey and T. Taniuti, ed. (New York: Academic Press, 1966).

This is probably the most widely referred to paper in the field. They derive the energy principle and give examples of how it can be applied for both the ideal MHD and the double adiabatic model ($p_\parallel \neq p_\perp$). Lagrangian coordinates are used for the derivation.

Von K. Hain, R. Lüst, and A. Schlüter, *Zeitschrift für Naturforschung*, *12A*, 833–841 (1957).

This paper is not widely read, partly because it has never been translated from German and partly because the authors use tensor notation. Their derivation of the energy principle is conceptually simpler because they use Eulerian (fixed) coordinates, as in this chapter, and because they retain the velocity as the dependent variable rather than transforming to the displacement ξ. They write expressions for δW in curvilinear coordinates, and they derive an upper bound on the possible MHD growth rates.

Other references cited in the text are

B. B. Kadomtsev, *Reviews of Plasma Physics*, 2, 153–199 (New York: Consultants Bureau, 1966).

R. M. Kulsrud, in *Advanced Plasma Theory*, Int. School of Physics Course, 25, Varenna, 1962, M. N. Rosenbluth, ed. (New York: Academic Press, 1964).

G. O. Spies, *Phys. Fluids*, 17, 2019–2024 (1974).

G. Laval, C. Mercier, and R. Pellat, *Nuclear Fusion*, 5, 156–158 (1965).

G. Schmidt, *Physics of High Temperature Plasmas* (New York: Academic Press, 1966).

B. S. Liley, *Plasma Physics*, 4, 325–328 (1962).

H. P. Furth, J. Killeen, M. N. Rosenbluth, and B. Coppi, Culham IAEA Conf., 1, 103–126 (1965).

J. M. Greene and J. L. Johnson, *Plasma Physics*, 10, 729–745 (1968).

G. Bateman, W. Schneider, and W. Grossmann, *Nuclear Fusion*, 14, 669–683 (1974).

A. Sykes and J. A. Wesson, *Nuclear Fusion*, 14, 645–648 (1974).

W. Grossmann and S. Ortolani, Max-Planck-Inst. für Plasmaphysik report IPP 1/132 (1973).

R. C. Grimm, J. M. Greene, and J. L. Johnson, in *Methods in Computational Physics*, J. Killeen, ed., 16, 253–280 (New York: Academic Press, 1976).

D. Berger, L. C. Bernard, R. Gruber, and F. Troyon, Berchtesgaden IAEA Conf., 2, 411–421 (1976).

L. S. Solov'ev, *Reviews of Plasma Physics*, 6, 239–331 (New York: Consultants Bureau, 1976).

D. Lortz, *Nuclear Fusion*, 13, 817–819 (1973).

6
Circular Cylinder Instabilities

There are two generic driving mechanisms for MHD instabilities in a circular cylinder. Pressure gradients together with the curvature of the magnetic field lines drive the so-called interchange instability, which is very similar to the Rayleigh-Taylor instability studied in chapter 3. This *interchange instability*, which causes one part of the plasma to try to change places with another part, is generally associated with some tolerable level of unstable activity localized to a region inside the plasma. The second type of instability, driven by current parallel to the magnetic field, is generally called a *kink instability*. Extreme examples of this type of instability, found in free-boundary plasmas surrounded by a vacuum region, can result in large distortions that are experimentally observed to carry the plasma to the wall. Of course, real MHD instabilities are driven by a combination of current, pressure gradient, curvature, etc. *Interchange* and *kink* are only loose categories representing a generalization of intuition based on extreme examples.

The circular cylinder geometry is a good place to begin the study of instabilities in confined plasmas—it is a step between plane slab and toroidal geometries. Circular cylindrical plasmas that are made in the laboratory by shock heating are generally called screw pinches because of the helical shape of the magnetic field. Alternatively, they are called Z-pinches when most of the plasma current is longitudinal, and Θ-pinches when most of the current is poloidal.

Circular Cylinder Instabilities

6.1 EQUILIBRIUM

In a straight circular cylinder all equilibrium quantities are functions of only the radius from the magnetic axis at the center of the cylinder. Using cylindrical coordinates (r, θ, z), the equilibrium force balance equation can be written

$$\frac{\partial p}{\partial r} = J_\theta B_z - J_z B_\theta \tag{6.1.1}$$

where

$$J_z = \frac{1}{\mu} \frac{1}{r} \frac{\partial}{\partial r} r B_\theta \tag{6.1.2}$$

and

$$J_\theta = -\frac{1}{\mu} \frac{\partial}{\partial r} B_z. \tag{6.1.3}$$

These equations can be combined to write

$$\frac{\partial}{\partial r}\left(p + \frac{1}{2\mu}B_z^2\right) + \frac{1}{\mu r} B_\theta \frac{\partial}{\partial r}(r B_\theta) = 0. \tag{6.1.4}$$

Alternatively, the equilibrium can be calculated from a stream function ψ determined by

$$\mu J_z = -\nabla^2 \psi = \mu p'(\psi) + B_z B_z'(\psi) \tag{6.1.5}$$

where

$$B_\theta = -\frac{\partial \psi}{\partial r}. \tag{6.1.6}$$

It is usually most convenient to specify the profiles $J_z(r)$ and $B_z(r)$ and then integrate (6.1.2) and (6.1.4) to determine $B_\theta(r)$ and then $p(r)$.

The q-value over a given length L of a circular cylinder is given exactly by

$$q(r) = \frac{kr B_z(r)}{B_\theta(r)}, \quad k \equiv 2\pi/L. \tag{6.1.7}$$

At any radius the q-value depends upon only the total amount of longitudinal current within that radius—independent of how that current is distributed. In practical units the q-value is given approximately by the formula

$$q(r) = 5k[\text{cm}^{-1}]r^2[\text{cm}]B_z(r)[\text{kG}]/I_z(r)[\text{kA}] \tag{6.1.8}$$

where $I_z(r)$ is the total longitudinal current within radius r. Taking the limit $r \to 0$, the q-value at the center of the cylinder depends upon only the current *density* and B_z where

$$q(r=0) = kB_z(r=0) \left/ \frac{\mu}{2} J_z(r=0) \right. . \tag{6.1.9}$$

If the current density is centrally peaked, the q-value is smallest at $r = 0$ and increases monotonically with radius. If J_z and B_z are uniform, the q-value is also uniform (zero shear) within the plasma. Without reversed currents a vacuum region yields the largest shear.

Question 6.1.1
It will be seen that a steep radial dependence of the q-value (high shear) has a strong effect on MHD instabilities. What is the q-value profile when the equilibrium is specified by $J_z = J_0(1 - r^2/a^2)^\nu$, $\nu > 0$, $B_z = 1$? In particular, what is the ratio of the q-value at the edge to the q-value at the center? An example of an experimentally measured profile is $J_z = J_0(1 - r^2/a^2)^4$ and $\rho = \rho_0(1 - r^2/a^2)^{1.7}$.

Question 6.1.2
Suppose the poloidal flux and the poloidal magnetic field at the edge of the plasma remain constant for some period of time. Does this imply that the current profile must have remained constant?

6.2 PHYSICAL PICTURE OF CURRENT DRIVEN INSTABILITIES

In order to gain some intuitive feeling, a number of physical pictures will be developed in this section for current driven instabilities in a free-boundary circular cylindrical plasma. These instabilities will be classified according to their poloidal mode number $e^{im\theta}$.

$m=0$ Sausage Instability

Consider a plasma with no longitudinal B_z field—only a poloidal B_θ field produced by a longitudinal current density J_z within the plasma. Suppose a poloidally symmetric radial perturbation constricts the plasma column in one place and makes it bulge out in other places, as illustrated in fig. 6.1. Since the same total current flows through the constricted area, the B_θ field at the plasma surface is increased there, and

Circular Cylinder Instabilities

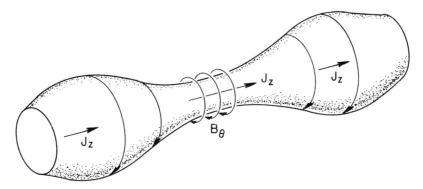

Fig. 6.1
$m = 0$ sausage instability.

the increased magnetic pressure makes the constriction contract further. The plasma pressure, however, does not increase substantially in the constricted region because the plasma is able to squeeze out into the bulged region. This $m = 0$ instability is inhibited when a longitudinal B_z field is present because the perturbation is forced to compress this field.

Question 6.2.1 (Kadomtsev, 1963)
Is there any profile that is stable to the sausage instability even without a B_z field? Evaluate the potential energy in the form given by (5.4.1) for a perturbation of the form $\boldsymbol{\xi} = \xi_r \hat{\mathbf{r}} + \xi_z \hat{\mathbf{z}}$. Then reduce the integrand of δW to a quadratic form in $\boldsymbol{\nabla} \cdot \boldsymbol{\xi}$ and ξ_r. From this determine the equilibrium condition needed for $\delta W > 0$.

$m = 1$ Kink Mode

The addition of a small longitudinal field stabilizes the sausage instability, but it destabilizes a new instability in which the perturbation looks like a corkscrew. This will be demonstrated with two simple models.

In the first model the plasma column is replaced with a thin wire carrying current I imbedded in a longitudinal magnetic field B_z. The wire is given a helical perturbation, shown in fig. 6.2. In the equilibrium state there is no force from $\mathbf{I} \times B_z \hat{\mathbf{z}}$, but in the perturbed state the outward force $\mathbf{I} \times B_z \hat{\mathbf{z}}$ accelerates the instability. In this case it is the perturbed current crossed with the equilibrium magnetic field that drives the instability.

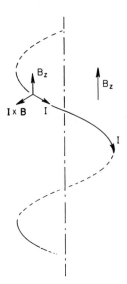

Fig. 6.2
$m=1$ kink instability in a thin plasma or wire.

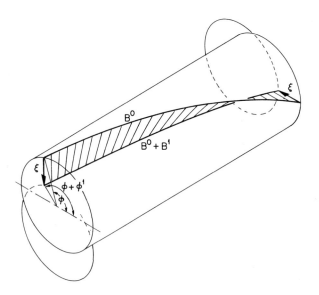

Fig. 6.3
$m=1$ kink instability in a fat plasma illustrating the Kruskal-Shafranov criterion. After J. L. Johnson, C. R. Oberman, R. M. Kulsrud, and E. A. Frieman, UN Geneva Conference, *31*, 198 (1958).

Question 6.2.2
What is the growth rate for this $m=1$ kink in a thin wire?

In the second model we consider a fat plasma column given a helical shift with rigid cross section. Consider two cross sections one quarter of a wavelength apart. If the equilibrium magnetic field subtends an angle $\varphi° > 90°$ around the magnetic axis as we move from one cross section to another, the perturbed magnetic field line will then subtend a larger angle $\varphi° + \varphi^1 > \varphi°$ around the original equilibrium magnetic axis on the side of the column where the perturbation is pushing inward, as shown in fig. 6.3. In order to traverse this larger angle, the poloidal component of the magnetic field must be increased by the perturbation. The added magnetic pressure then accelerates the perturbation inward on this side of the column. Note that the equilibrium condition $\varphi° > 90°$ is equivalent to $\iota > 1$ for the rotational transform or $q < 1$ for the q-value at the edge of the plasma.

Repeating the same argument for an equilibrium with $\varphi° < 90°$ (or $q > 1$), we see that the perturbation decreases the angle φ and therefore decreases the poloidal magnetic field and the magnetic pressure from $J_z^0 \times B_{\text{pol}}^1$ on the side being pushed in. Hence the perturbation is opposed.

This simple model indicates that the $m=1$ kink mode is unstable if $q_{\text{edge}} < 1$ and stable if $q_{\text{edge}} > 1$. The stability criterion $q_{\text{edge}} > 1$ is called the *Kruskal-Shafranov stability criterion*. It is probably the most important stability criterion for tokamaks and pinches. It constitutes a large part of the rationale for building tokamaks with strong toroidal magnetic fields. Given the strongest toroidal field we can afford, the Kruskal-Shafranov criterion sets an upper bound on the amount of toroidal current we can drive through the plasma column. The stability criterion $q > 1$ will appear many times and in many forms in the chapters to follow.

This physical picture of the $m=1$ kink instability was taken from Johnson, Oberman, Kulsrud, and Frieman (1958, pp. 207–208).

Question 6.2.3
Consider a uniform current cylindrical plasma with a Cartesian coordinate system ($B_x = -J_{z0}y/2$, $B_y = J_{z0}x/2$, $B_z = B_{z0}$). Given an $m=1$ perturbation $\mathbf{v}^1 = (iv_y, v_y, v_z(x - iy)/a)e^{\gamma t + im\theta - ikz}$ and $\mathbf{B} = (b_x, -ib_x, b_z(y + ix)/a)e^{\gamma t + im\theta - ikz}$, it is possible to find a consistent relation between the constants v_y, b_x, v_z, b_z to lowest order in the aspect ratio ($ka \ll 1$) from the original linearized MHD equations (5.1.1) and (5.1.2),

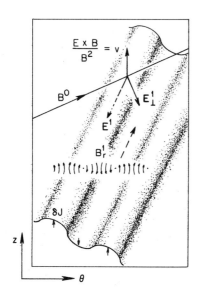

Fig. 6.4
Sequence of steps for an $m \geqslant 2$ kink instability. After R. S. Lowder and K. I. Thomassen, *Phys. Fluids*, *16*, 1497 (1973).

neglecting pressure completely. From which equation does the Kruskal-Shafranov stability condition $\frac{1}{2} J_{z0} < k B_z$ come? Which term in the equation of motion drives the instability?

$m \geqslant 2$ Kink Modes

Suppose we unwrap the surface of the plasma and flatten it out to form the rectangle shown in fig. 6.4. This figure should be read from the bottom up in order to illustrate the sequence of events that drive the instability. The radial part of the perturbation has a sinusoidal dependence $\sin(m\theta)$, shifting the plasma into and out of the surface. If the equilibrium current density has a sharp radial gradient at the plasma surface, the perturbation lifts part of this current up out of the surface according to $\delta \mathbf{J} \simeq \boldsymbol{\xi} \cdot \nabla \mathbf{J}^0$. This produces a perturbed radial magnetic field, B_r^1, with an alternating direction around θ which, in turn, induces an electric field parallel to the ridges of the perturbation

$$-\oint d\mathbf{l} \cdot \mathbf{E}^1 = \frac{\partial}{\partial t} \int d\mathbf{S} \cdot \mathbf{B}_r^1. \tag{6.2.1}$$

The plasma short circuits that component of the perturbed electric field that is parallel to the equilibrium magnetic field. The remainder of the perturbed electric field drives an **E** × **B** drift into or out of the surface. If the equilibrium magnetic field wraps around the cylinder faster than the ridges of the perturbation, shown in fig. 6.4, the resulting **E** × **B** drift enhances the perturbation. Otherwise it opposes the perturbation. Also, if the angle between the magnetic field and the ridges of the perturbation is too great, the instability is suppressed because it tends to bend the magnetic field too much—just as in the Rayleigh-Taylor instability with shear.

Here, the angle between the magnetic field lines and the perturbation ridges is critical. This model predicts instability if $mq < n$ and stability if $mq > n$ at the surface of the plasma, where m and n are the poloidal and longitudinal mode numbers. If the magnetic field lines do not form a simple helix, as in more complicated geometries, one would expect the perturbation to be somewhat deformed and grow preferentially where the angle relative to the magnetic field is most favorable.

This model was devised by Lowder and Thomassen (1973).

Question 6.2.4
Can you derive $\delta \mathbf{J} \simeq \boldsymbol{\xi} \cdot \nabla \mathbf{J}^0$? What other terms contribute to $\delta \mathbf{J}$? Would there still be a kink instability if $J^0(r)$ and $\partial J^0(r)/\partial r$ both go smoothly to zero at the edge of the plasma?

Question 6.2.5
Suppose there is a nearby mode-rational surface $q = m/n$ where the equilibrium magnetic field is parallel to the perturbation ridges. Does it make any difference to the instability if this mode rational surface is in a vacuum or imbedded in a perfectly conducting fluid?

6.3 THE 1-D EIGENVALUE EQUATION

For a circular cylinder the coefficients in the linearized MHD equations are independent of time and also independent of the coordinates θ and z. It follows that Fourier harmonics in these coordinates evolve independently and, therefore, we may consider one Fourier harmonic at a time. The perturbation may then be written in the form

$$\boldsymbol{\xi}(r,t) = \boldsymbol{\xi}(r) e^{\gamma t + i(m\theta - kz)}. \tag{6.3.1}$$

Since this Fourier harmonic is a function of radius alone, the problem has been reduced to one dimension.

All but two of the perturbation variables may be eliminated from the MHD equations by algebraic means. There are two key steps in this algebraic reduction. One step is the definition of the total perturbed pressure p^* as the sum of thermodynamic plus magnetic pressure

$$p^* \equiv p^1 + \mathbf{B}^0 \cdot \mathbf{B}^1/\mu. \tag{6.3.2}$$

The other key step is the derivation of the compression term

$$\nabla \cdot \xi = \frac{-\rho\gamma^2(p^* - (2/\mu r)B_\theta^2 \xi_r)}{(\rho\gamma^2 + F^2/\mu)\Gamma p^0 + \rho\gamma^2(B_\theta^2 + B_z^2)/\mu} \tag{6.3.3}$$

where F stands for the scalar product of the wave vector and the equilibrium magnetic field

$$F \equiv \mathbf{K} \cdot \mathbf{B} \equiv \frac{m}{r} B_\theta - k B_z = \frac{B_\theta}{r}(m\text{-}nq). \tag{6.3.4}$$

After algebraic reduction it is found that the variables $r\xi_r$ and p^* satisfy a pair of ordinary differential equations

$$(\rho\gamma^2 + F^2/\mu)r\frac{\partial}{\partial r}(r\xi_r) = C_{11} r\xi_r + C_{12} p^* \tag{6.3.5}$$

$$(\rho\gamma^2 + F^2/\mu)r\frac{\partial}{\partial r} p^* = C_{21} r\xi_r - C_{11} p^* \tag{6.3.6}$$

where

$$C_{11} = 2(mFB_\theta/r - B_\theta^2 \Lambda)/\mu$$

$$C_{12} = r^2 \Lambda - m^2 - k^2 r^2$$

$$C_{21} = -(\rho\gamma^2 + F^2/\mu)^2 + 2(\rho\gamma^2 + F^2/\mu) B_\theta \frac{d}{dr}\left(\frac{B_\theta}{r}\right)/\mu$$

$$+ 4\left(\frac{B_\theta}{r}\right)^2 (F^2/\mu - B_\theta^2 \Lambda)/\mu$$

$$\Lambda = -\rho^2\gamma^4/[(\rho\gamma^2 + F^2/\mu)\Gamma p^0 + \rho\gamma^2(B_\theta^2 + B_z^2)/\mu].$$

The boundary condition at $r = 0$ is found by making a power series expansion of these equations near $r = 0$ and solving the resulting indicial equation to show that

$$\xi_r \to r^{m-1} \tag{6.3.6}$$

and

$$p^* \to \frac{1}{m}\left(\frac{2 B_\theta}{\mu r} F - \rho\gamma^2 - F^2/\mu\right) r\xi_r. \tag{6.3.7}$$

Circular Cylinder Instabilities

If the plasma extends out to a rigid wall, the boundary condition there is

$$\xi_r(r = a) = 0. \tag{6.3.8}$$

As in any linearized problem, the amplitude of the solution is arbitrary.
All the other perturbed variables can then be expressed in terms of $r\xi_r$ and p^*

$$i\xi_\theta = \left(\frac{m}{r}p^* - \frac{2B_\theta}{\mu r}F\xi_r + B_\theta F \nabla \cdot \xi/\mu\right) \bigg/ \left(\rho\gamma^2 + F^2/\mu\right) \tag{6.3.9}$$

$$i\xi_z = (-kp^* + FB_z \nabla \cdot \xi/\mu)/(\rho\gamma^2 + F^2/\mu) \tag{6.3.10}$$

$$iB_r^1 = -F\xi_r \tag{6.3.11}$$

$$B_\theta^1 = Fi\xi_\theta - B_\theta \nabla \cdot \xi - r\xi_r \frac{d}{dr}\left(\frac{B_\theta}{r}\right) \tag{6.3.12}$$

$$B_z^1 = Fi\xi_z - B_z \nabla \cdot \xi - \xi_r \frac{d}{dr}B_z. \tag{6.3.13}$$

Question 6.3.1
For stable oscillations, where γ^2 is negative, what is the nature of the eigenfunction near the radius where $F^2 = -\rho\gamma^2\mu$?

Question 6.3.2
Why is the radial magnetic field perturbation B_r^1, given by (6.3.11), zero at the mode rational surface where $F = 0$? Does it make any difference if the mode rational surface is in a perfectly conducting plasma or a resistive region?

If there is a vacuum region surrounding the plasma, the components of each Fourier harmonic of the perturbed magnetic field there have the form

$$-iB_r^1 = k[C_1 I_m'(kr) + C_2 K_m'(kr)] \tag{6.3.14}$$

$$B_\theta^1 = -\frac{m}{r}[C_1 I_m(kr) + C_2 K_m(kr)] \tag{6.3.15}$$

$$B_z^1 = k[C_1 I_m(kr) + C_2 K_m(kr)] \tag{6.3.16}$$

where I_m and K_m are hyperbolic Bessel functions. By forcing this perturbed magnetic field to satisfy the boundary conditions $B_r^1 = 0$ at the radius of a perfectly conducting wall (r_{wall}), and by making B_r^1 continuous at the plasma-vacuum interface, the constants C_1 and C_2

can be expressed in terms of the radial displacement ξ_a at the edge of the plasma

$$C_1 = \xi_a F(a) K'_m(kr_{\text{wall}})/k\Delta$$
$$C_2 = -\xi_a F(a) I'_m(kr_{\text{wall}})/k\Delta \qquad (6.3.17)$$

where

$$\Delta \equiv I'_m(ka) K'_m(kr_{\text{wall}}) - I'_m(kr_{\text{wall}}) K'_m(ka).$$

Finally, imposing the condition that p^* be continuous at the plasma-vacuum interface yields the relation

$$\xi_a = \left(\frac{k\mu}{F^2}\right) \frac{I'_m(ka) K'_m(kr_{\text{wall}}) - K'_m(ka) I'_m(kr_{\text{wall}})}{K_m(ka) I'_m(kr_{\text{wall}}) - I_m(ka) K'_m(kr_{\text{wall}})} p^*(a). \qquad (6.3.18)$$

It can be seen that the addition of a vacuum region is straightforward.

In general, it can be proven that the unstable eigenvalues of (6.3.5) and (6.3.6) behave like the eigenvalues of the Sturm-Liouville problem (see, for example, Goedbloed and Sakanaka (1974)). Using this fact, a simple iteration scheme can be used to compute the solutions of (6.3.5) and (6.3.6) Start with any guess for the eigenvalue γ and integrate the equations out to the edge of the plasma. If the value of ξ_r is too large there, reduce γ and compute again; if ξ_r is too small, increase γ. The same idea can be generalized to allow for a prescribed number of radial nodes where $\xi_r = 0$ within the plasma, and the accuracy of the result can be verified by integrating backwards to the origin. Computer codes using this method, called "shooting codes" are in widespread use. The growth rate curves and eigenfunctions illustrated in this chapter were computed using such a code.

6.4 THE 1-D ENERGY PRINCIPLE

While the eigenvalue equations are well suited for computational work, it is more convenient to use a one-dimensional form of the potential energy for analytic work. This form can be derived by partially minimizing δW with respect to the components of $\boldsymbol{\xi}$ in the flux surfaces. W. A. Newcomb (1960) goes through the details of this derivation as well as giving an extensive discussion of the utility of the 1-D energy principle. After these partial minimizations, the energy principle is useful only for distinguishing stable from unstable equilibria—not for determining eigenfunctions or growth rates—as explained in chapter 5. Starting with the general form of δW within the plasma given by (5.4.2) the

Circular Cylinder Instabilities

1-D form of δW is

$$\delta W_F = \frac{\pi}{2\mu} \int_0^a dr r \left\{ \frac{r^2 B_\theta^2}{m^2 + k^2 r^2} [(nq - m)\xi_r' - (nq + m)\xi_r/r]^2 \right.$$

$$\left. + [r^2 B_\theta^2 (nq - m)^2 - 2B_\theta (rB_\theta)'] \xi_r^2/r^2 \right\}. \quad (6.4.1)$$

One of the conditions for minimization is $\nabla \cdot \boldsymbol{\xi} = 0$.

If there is a vacuum region around the plasma, the potential energy due to the perturbed magnetic fields in the vacuum region (5.4.5) should be added to (6.4.1). Since these perturbed magnetic fields are completely determined by (6.3.14)–(6.3.18), δW_{vac} can be evaluated directly in terms of the displacement at the edge of the plasma. In the limit of a long narrow cylinder, $ka \lesssim kr_{wall} \ll 1$, δW_{vac} is

$$\delta W_{vac} = \frac{\pi}{2\mu} \xi_a^2 B_{\theta a}^2 \frac{(m - nq)^2}{m} \frac{1 + (a/r_{wall})^{2m}}{1 - (a/r_{wall})^{2m}}. \quad (6.4.2)$$

This always makes a positive contribution to the potential energy.

Question 6.4.1
How can free-boundary instabilities grow faster than fixed-boundary instabilities if the potential energy from the vacuum is always positive (stabilizing)?

The term in (6.4.1) involving $\xi_r \xi_r'$ can be integrated by parts to yield the most useful expression for the potential energy

$$W = W_f + W_a + W_{vac} \quad (6.4.3)$$

$$W_f = \frac{\pi}{2} \int_0^a dr [f(\xi_r')^2 + g\xi_r^2] \quad (6.4.4)$$

$$f = rB_\theta^2 (m - nq)^2 / \mu (m^2 + k^2 r^2)$$

$$g = \frac{2k^2 r^2}{m^2 + k^2 r^2} \frac{dp}{dr} + \frac{B_\theta^2}{\mu r} \left[(m - nq)^2 \frac{m^2 - 1 + k^2 r^2}{m^2 + k^2 r^2} \right.$$

$$\left. + \frac{2k^2 r^2 ((nq)^2 - m^2)}{(m^2 + k^2 r^2)^2} \right]$$

$$W_a = \frac{\pi}{2\mu} \xi_a^2 B_{\theta a}^2 \left. \frac{(m - nq)^2 - 2m(m - nq)}{m^2 + k^2 r^2} \right|_{r=a} \quad (6.4.5)$$

An additional contribution (5.4.6) must be added if there is a surface current in the equilibrium.

In the limit of a long thin cylinder, called the large aspect ratio limit, $k^2r^2 \ll 1$, the driving term within the plasma can be simplified to

$$g = \begin{cases} \dfrac{B_\theta^2}{\mu r m^2}\left[(m^2-1)(nq-m)^2 + \dfrac{2\mu k^2 r^2}{B_\theta^2}r\dfrac{dp}{dr}\right] & (6.4.6) \\ \text{for } m \geqslant 2 \\ \dfrac{B_\theta^2}{\mu r}k^2 r^2\left[3(nq-1)^2 + 4(nq-1) + \dfrac{2\mu}{B_\theta^2}r\dfrac{dp}{dr}\right] & (6.4.7) \\ \text{for } m=1 \end{cases}$$

The $m=1$ mode is a special case because the zero-order terms in the $(kr)^2$ expansion cancel out.

6.5 FIXED-BOUNDARY INSTABILITIES

For fixed boundary or internal instabilities, there is no radial perturbation $\xi_r = 0$ at the edge of the plasma and so there is no flow of energy, mass, or flux across the boundary $r = a$. Only the internal contribution δW_f is used for the energy principle, with no surface or vacuum contribution. Poloidal fields and flows, however, are still allowed at the edge of the plasma.

Internal $m=1$ mode

The strategy here is to construct a test function for $\xi_r(r)$ which is concentrated in the region where the $g\xi^2$ term in (6.4.4) is most negative without making the positive $f(\xi')^2$ term too large. As long as $p'(r)$ is negative (pressure decreasing away from the axis), it suffices to use a step function for ξ_r, which is positive wherever $q(r) < 1$ and zero elsewhere

$$\xi_r(r) = \begin{cases} \xi_a & q(r) < 1 \\ 0 & q(r) > 1, \end{cases} \quad (6.5.1)$$

to demonstrate instability. In the large aspect ratio limit $k^2r^2 \ll 1$, this step function is a very good approximation to the function that minimizes δW_f. If any test function with radial derivative substantially different from zero were used in the region where $(nq-1)^2 \neq 0$, the first term in (6.4.4) would dominate δW_f with a positive (stabilizing) contribution.

Question 6.5.1

The solution of the Euler equation

$$(f\xi')' - g\xi = 0 \tag{6.5.2}$$

is supposed to minimize δW_f given by (6.4.4). Substituting (6.5.2) back into (6.4.4) yields

$$\delta W_f = \int dr \left[f\xi'^2 + (f\xi')' \xi \right] = 0$$

if $\xi\xi' = 0$ at $r=0$ and $r=a$. We have just demonstrated a test function (6.5.1), however, that makes δW_f *negative* is $q(0) < 1$. Why doesn't the Euler equation find the proper minimum for δW_f?

The real flow pattern for the $m=1$ eigenfunction is a pair of vortex cells with a nearly uniform flow across the center of the plasma, and a large return flow concentrated at the mode rational surface $q(r) = 1$. Flow patterns are illustrated in fig. 6.5 for a uniform current equilibrium where the structure of the modes is less singular. There is very little flow down the cylinder. The perturbed magnetic field also consists of a pair of vortex patterns, 90° out of phase with the velocity vortex cells. The effect of the flow pattern is to convect the plasma away from the center of the cylinder within the mode rational surface. The nonlinear development of this convection will be discussed in chapter 9.

A simple method for estimating the growth rate of this instability was suggested by Rosenbluth, Dagazian, and Rutherford (1973). In this derivation, it is assumed that ξ_z and $\nabla \cdot \xi$ can be neglected when $kr \ll 1$ and $|B_\theta| \ll |B_z|$. High shear is also assumed, so that the step function (6.5.1) is a good approximation.

From these assumptions it follows that the kinetic energy per unit length of the cylinder has the form

$$\begin{aligned} KE &= \pi \int dr\, r\, \gamma^2 \rho^0 (\xi_r^2 + \xi_\theta^2) \\ &= \pi \int dr\, r\, \gamma^2 \rho^0 \left[\xi_r^2 + \left(\frac{d}{dr} r\xi_r \right)^2 \right]. \end{aligned} \tag{6.5.3}$$

Minimizing the sum of kinetic and potential energy with respect to a variational function $\delta\xi_r$, which is fixed at the wall and at the origin, results in the Euler equation

$$[(2\gamma^2 r^3 \rho + f)\xi_r']' - (g - 2r^2\gamma^2\rho')\xi_r = 0. \tag{6.5.4}$$

It can be shown a posteriori that the $2r^2\gamma^2\rho'$ term can be neglected in

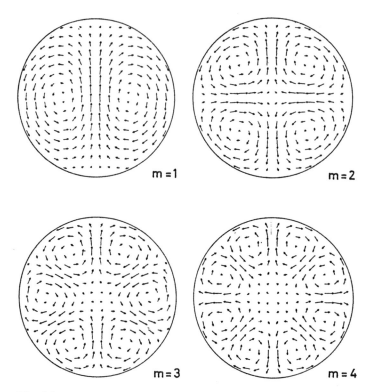

Fig. 6.5
Fixed boundary instabilities in a uniform current equilibrium with $ka=1$. From G. Bateman, W. Schneider, and W. Grossmann, *Nuclear Fusion*, 14, 669 (1974).

comparison to g. There are a number of ways to approximate the solution to (6.5.4). The step function (6.5.1) is a solution of (6.5.4) to lowest order in $(kr)^2$. Using this step function, (6.5.4) can be integrated once to get an approximation for the small slope of ξ away from the mode rational surface

$$\xi'_r = \begin{cases} \dfrac{\xi_a}{f + 2\gamma^2 r^3 \rho} \displaystyle\int_0^r dr g & 0 < r < r_s \\[2ex] \dfrac{\text{const.}}{f + 2\gamma^2 r^3 \rho} & r_s < r < a. \end{cases} \quad (6.5.5)$$

Finally, an approximate solution for ξ_r can be found in the immediate neighborhood of the mode rational surface $q(r_s) = 1$

$$|r - r_s| \equiv |x| \ll 1 \quad (6.5.6)$$

Circular Cylinder Instabilities

by neglecting the driving term $g\xi_r$ and using the approximation

$$f \simeq r^3(k \cdot B)^2 \simeq r_s^3|(k \cdot B)'|^2 x^2. \tag{6.5.7}$$

The resulting equation can be integrated and the solution can be matched to the step function (6.5.1) far from the mode rational surface

$$\xi_r \simeq \int dx \frac{\text{const.}}{2\gamma^2 \rho(r_s) + |(k \cdot B)'|^2 x^2/\mu} \tag{6.5.8}$$

$$\simeq \tfrac{1}{2}\xi_a \left\{ 1 - \frac{2}{\pi} \tan^{-1}[x|(k \cdot B)'|_{r_s}/\gamma\sqrt{2\rho\mu}] \right\}.$$

In order to match the radial derivative of this solution to (6.5.5) in the region where $r < r_s$, the following condition must be satisfied

$$\gamma \simeq \frac{\pi}{\sqrt{2\rho(r_s)/\mu}|(k \cdot B)'|_{r_s} r_s^3} \int_0^{r_s} dr(-g). \tag{6.5.9}$$

This formula for the growth rate works reasonably well when the mode rational surface r_s is in a high shear region well away from the origin. But (6.5.9) breaks down completely, and computational results must be used when the mode rational surface is close to the origin ($r_s \to 0$), as illustrated in fig. 6.6.

$m \geq 2$ Instabilities

For $m \geq 2$ modes, the driving term g in (6.4.4) is negative over only a narrow range of q-values

$$|nq(r) - m| \lesssim \frac{\mu kr}{B_\theta(r)} \left[\frac{-2rp'(r)}{m^2 - 1} \right]^{1/2} \tag{6.5.10}$$

For equilibria with high shear (q varying rapidly with r), the instability is driven only in a thin shell around the mode rational surface. It follows that $m \geq 2$ modes are localized near their mode rational surfaces and their growth rates are low.

Computational results indicate that these higher mode numbers are completely suppressed by high shear and large aspect ratio. Growth rates get progressively smaller for higher mode numbers and longer wavelengths, as illustrated in fig. (6.6).

Instabilities With Many Radial Nodes

Each eigenfunction in a straight cylindrical plasma can be characterized by three mode numbers (k, m, n). The longitudinal wave number

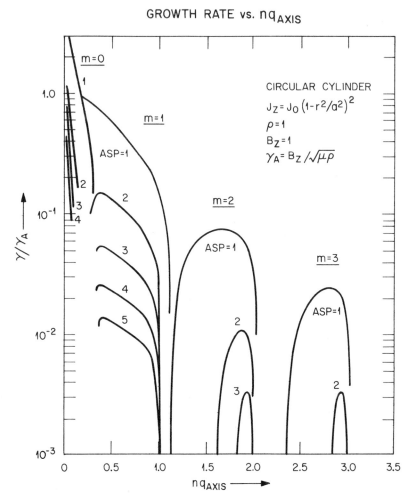

Fig. 6.6
Growth rates of fixed boundary instabilities in a circular cylinder with moderate shear and different aspect ratios.

Circular Cylinder Instabilities

k, which is inversely proportional to the longitudinal wavelength ($k = 2\pi/L$), and the poloidal mode number m have already been considered. Now we shall consider eigenfunctions with different radial mode numbers n, where n is the number of nodal lines ($\xi_r = 0$) encircling the origin. For any given mode numbers k and m, there are an infinite number of radial modes corresponding to stable oscillations, and there may be a finite or an infinite number of radial modes corresponding to different unstable eigenfunctions.

Goedbloed and Sakanaka (1973) proved that, for any given m and k, unstable eigenfunctions with successively more radial nodes have progressively smaller growth rates. The eigenfunction with no radial nodes ($n = 0$) is *always* the most unstable, if it is unstable at all. This theorem is true for straight cylinders but is not true for bumpy cylinders and certain closed field line toroidal configurations.

A series of $m=1$ unstable eigenfunctions with successively more radial nodes is shown in fig. 6.7. It can be seen that the radial oscillations concentrate in the neighborhood of the mode rational surface. If any mode localized near the mode rational surface is unstable, then the corresponding large-scale instability with no radial nodes is unstable, too. Localized instabilities must be stabilized before large-scale instabilities can be stabilized. For given poloidal and longitudinal mode

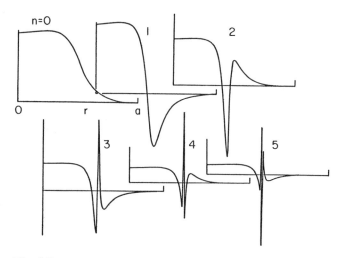

Fig. 6.7
Radial perturbation $\xi_r(r)$ for fixed-boundary $m=1$ instabilities with an increasing number of radial nodes. From J. P. Goedbloed and P. H. Sakanaka, *Phys. Fluids*, *17*, 908 (1974).

numbers, any local stability criterion is necessary for large-scale stability as well.

A stability criterion for modes localized arbitrarily close to a mode rational surface was derived by Bergen Suydam (1958). (The last name is pronounced as you would say "it's a damn shame.") Suydam started with the Euler equation (6.5.2) for minimizing the internal potential energy δW_f (6.4.4). In order to look for solutions which are localized arbitrarily close to a mode rational surface at $r = r_s$, a Taylor series expansion of the Euler equation is made in the variable $x = r - r_s$. To lowest order in this expansion, the Euler equation becomes

$$(x^2 \xi_r')' + D_s \xi_r = 0 \tag{6.5.11}$$

where

$$D_s \equiv -\frac{q^2}{q'^2} \frac{2\mu p'}{B_z^2 r_s} \tag{6.5.12}$$

is the driving term for instabilities. Solutions of this equation have the form

$$\xi_r = x^s \tag{6.5.13}$$

for values of s satisfying the indicial equation

$$s(s + 1) + D_s = 0$$
$$s = \tfrac{1}{2}[-1 \pm (1 - 4D_s)^{1/2}]. \tag{6.5.14}$$

Since the real part of s is negative, the solutions (6.5.13) are singular at the mode rational surface $x = 0$. If these solutions are truncated arbitrarily close to the mode rational surface, we have perfectly valid test functions that are close to the minimizing eigenfunctions almost everywhere and can be substituted back into δW_f to test for stability. The perturbation is stable ($\delta W_f > 0$) if $D_s < \tfrac{1}{4}$ and unstable if $D_s > \tfrac{1}{4}$. The unstable eigenfunctions have the form of radial oscillations which oscillate infinitely rapidly and grow to infinite amplitude as $x \to 0$

$$\xi_r = x^{-1/2} \cos[\tfrac{1}{2}(4D_s - 1)^{1/2} \ln x + \phi]\{1 + \cdots\}. \tag{6.5.15}$$

The *Suydam stability criterion* $D_s < \tfrac{1}{4}$ is usually written

$$\left(\frac{q'}{q}\right)^2 > \frac{-8\mu p'}{r B_z^2}. \tag{6.5.16}$$

It has been generalized by Shafranov (1970) in order to include finite aspect ratio effects in a straight cylinder. The method used to derive

Circular Cylinder Instabilities

the Suydam criterion will appear again in the derivations of the Mercier criterion for arbitrary toroidal shapes (section 7.2) and the resistive interchange stability criterion (section 10.4).

Question 6.5.2
A pressure gradient drives the Suydam instability and shear $(q'/q)^2$ tends to stabilize it. Is this shear stabilization effective near $r = 0$? Would it be effective if the pressure were flattened there? If the instability produces convection which flattens the pressure profile, would this be a self-stabilizing effect?

Question 6.5.3
In the expression for δW_f (6.4.4), ξ'_r appears only in a positive definite term. The test function for the Suydam instability oscillates rapidly in radius so that $f(\xi'_r)^2$ is large. How can this be an unstable eigenfunction?

6.6 FREE-BOUNDARY INSTABILITIES

When people speak about free-boundary plasmas, they are usually referring to a discontinuity between the perfectly conducting plasma and a vacuum region surrounding the plasma. In effect, there is a discontinuity in the resistivity profile—since high resistivity is the essence of a vacuum. The sharp distinction made in the literature (by Shafranov (1970), for example) between free-boundary instabilities, which deform the plasma surface, and fixed-boundary instabilities, which may drive only parallel flows at the plasma surface, appears to be an important distinction only in the context of broad current profiles. The distinction is quite clear for a uniform current profile, for example, where the growth rates of the free-boundary instabilities are much higher than the internal instabilities and their domains of instability are somewhat different (see, for example, Takeda et al. (1972)). However, the distinction virtually disappears for highly peaked current profiles where the higher m poloidal modes are stabilized by shear and the free-boundary $m=1$ instability appears as a natural extension of the internal $m=1$ instability as the q-value at the surface of the plasma becomes less than one and the corresponding q-values within the plasma are much less than one.

Three kinds of ideal MHD instabilities are discussed in a free-boundary cylindrical plasma.

The free-boundary $m=1$ instability is distinguished by the fact that it is unstable whenever $q < 1$ at the edge of the plasma (in a low beta plasma) regardless of the current profile. This Kruskal-Shafranov stability condition can be easily demonstrated by observing that the internal part of the potential energy (6.4.4) is negligible in comparison with the surface (6.4.5) and vacuum (6.4.2) parts in the large aspect ratio limit $k^2 r^2 \ll 1$. The condition $\delta W_a + \delta W_{\text{vac}} < 0$ is found for the domain of instability

$$\frac{a^2}{r_{\text{wall}}^2} < q_a < 1 \tag{6.6.1}$$

where a is the plasma radius, r_{wall} is the radius to a perfectly conducting wall, and q_a is the q-value at the edge of the plasma. The lower bound on the q-value is due to wall stabilization—where the image currents in the wall produce forces that oppose the helical form of the instability. The growth rate of the instability can be approximated by taking the perturbation to be a helical distortion of the plasma column made up of rigid shifts at each cross section with no longitudinal motion. The kinetic energy per unit wavelength λ reduces to

$$\text{KE} = \int d^3 x \rho \gamma^2 \xi^2 \simeq \langle \rho \rangle \gamma^2 \xi_a^2 \pi a^2 \lambda \tag{6.6.2}$$

where $\langle \rho \rangle$ is the space averaged mass density within the plasma column. Equating this kinetic energy with the loss of potential energy due to the perturbation, the growth rate is found to be

$$\gamma^2 = \frac{B_{\theta a}^2}{\mu \langle \rho \rangle a^2 \lambda} \left[1 - q_a - \frac{(1 - q_a)^2}{1 - a^2/r_{\text{wall}}^2} \right]. \tag{6.6.3}$$

The second type of free boundary instability, with $m \geq 2$, is strongly suppressed by shear like that found near the edge of a plasma with peaked current profile. This was demonstrated analytically by Shafranov (1970) using a specific equilibrium, and it is routinely observed in computational studies. It is quite sensitive to the amount of shear present.

The third type of free-boundary instability is called a *peeling instability* which occurs only when there is a discontinuity in the current density or one of its low-order derivatives at the edge of the plasma. As its name implies, this instability appears to peel off the surface of the plasma as a thin layer. The peeling instability was introduced by Frieman, Greene, Johnson, and Weimer (1973), and it is discussed in the review paper on linearized MHD instabilities by Wesson (1975).

Question 6.6.1

Suppose the vacuum region were replaced by a perfactly conducting plasma that carried no equilibrium current. How much would the free-boundary $m = 1$ mode be affected?

Question 6.6.2

Can an internal instability exist at the same time as a free-boundary instability? Can they be completely independent?

6.7 SUMMARY

It is best to use the eigenvalue equations (6.3.5) and (6.3.6) to determine the growth rates and spatial structure of MHD instabilities in a circular cylinder. Alternatively, the energy principle (6.4.2)–(6.4.5) may be used to distinguish stable from unstable equilibria.

Circular cylinder instabilities are categorized by their poloidal mode number m and their radial mode number n for a given longitudinal wave number k (6.3.1).

The $m = 1$ mode is unstable whenever $q < 1 + \mathcal{O}(ka)$ anywhere in the plasma with $p'(r)$ negative. The instability appears as a nearly rigid helical deformation of that part of the plasma within the $q = 1$ mode rational surface.

The $m \geqslant 2$ instabilities are localized near the $nq(r) = m$ mode rational surfaces and are stabilized by high shear at large aspect ratios.

The Suydam criterion (6.5.16) is a necessary condition for stability, derived using radially localized modes. These must be stabilized before larger scale modes can be stabilized.

6.8 REFERENCES

V. D. Shafranov, *Sov. Phys.—Tech. Phys.*, *15*, 175–183 (1970).

This paper was a turning point in the MHD literature—turning away from a preoccupation with the marginal points of localized instabilities. Shafranov concentrates on relevant equilibrium profiles and the effects of realistic assumptions, making estimates for the growth rates and spatial structure of large scale instabilities. It is an excellent review.

For a readable account of recent ideas on spectral theory and the computation of instabilities see

J. P. Goedbloed and P. H. Sakanaka, *Phys. Fluids*, *17*, 908–929 (1974).

Two classic papers on marginal point analysis are

W. A. Newcomb, *Ann. Phys.* (N.Y.), *10*, 232–267 (1960).

B. R. Suydam, IAEA Geneva Conf., *31*, 157–159 (1958).

For numerical methods and results refer to

R. J. Tayler, *Proc. Phys. Soc.* (London), *B70*, 1049–1063 (1957).

T. Takeda, Y. Shimomura, M. Ohta, and M. Yoshikawa, *Phys. Fluids, 15*, 2193–2201 (1972).

J. P. Goedbloed and H. J. L. Hagebeuk, *Phys. Fluids, 15*, 1090–1101 (1972).

W. Grossmann and S. Ortolani, Max-Planck-Inst. für Plasmaphysik report IPP 1/132 (1973).

A comparison between theory and experiment is given by

J. P. Freidberg, *Phys. Fluids, 13*, 1812–1818 (1970).

A clear paper on the stability of reversed field pinches, a subject not treated here, is provided by

D. C. Robinson, *Plasma Physics, 13*, 439–462 (1971).

Other papers cited in the text are

B. B. Kadomtsev, *Reviews of Plasma Physics, 2*, 153–199, New York: Consultants Bureau, 1966.

J. L. Johnson, C. R. Oberman, R. M. Kulsrud, and E. A. Frieman, UN Geneva Conf., *31*, 198–212 (1958).

R. S. Lowder and K. I. Thomassen, *Phys. Fluids, 16*, 1497–1500 (1973).

M. N. Rosenbluth, R. Y. Dagazian, and P. H. Rutherford, *Phys. Fluids, 16*, 1894–1902 (1973).

E. A. Frieman, J. M. Greene, J. L. Johnson, and K. E. Weimer, *Phys. Fluids, 16*, 1108–1125 (1973).

J. A. Wesson, Seventh European Conf. on Controlled Fusion and Plasma Physics, Lausanne, *2*, 102–118 (1975).

7
Toroidal Instabilities

The first part of this chapter is concerned with the Mercier stability criterion, which is a local stability criterion applicable to plasmas with arbitrary toroidal geometry and any beta. Greene and Johnson's (1962) derivation of the Mercier criterion will be given in considerable detail in section 7.2 because it uses mathematical techniques that are typical of the best analytic literature on MHD instabilities. A central part of this derivation is the use of flux coordinates, introduced in section 7.1. Some applications of the Mercier criterion to axisymmetric toroidal plasmas will be given in section 7.3.

The final section of this chapter provides a complete change of pace with a discussion of computational results on large scale MHD instabilities in tokamak equilibria. The effect of toroidicity on linear instabilities in low beta equilibria will be emphasized. High beta results are reserved for chapter 8 and nonlinear results for chapter 9.

7.1 FLUX COORDINATES

To be able to read most of the literature on the stability of toroidal systems, the reader must gain some facility with *flux coordinates*— curvilinear coordinate systems in which one of the coordinates is constant over each flux surface, and the other two coordinates form a grid within each flux surface. There are a wide variety of possible choices among these flux coordinates. Any unique labeling of the flux surfaces may be used for the "radial" coordinate, and the grid of "angle" coordinates within each surface may be deformed at will. It is possible to choose either orthogonal flux coordinates or coordinates in which both the magnetic and current field lines appear as straight lines (called Hamada coordinates), which are generally not orthogonal. This survey will start with general possibilities and work toward those specific choices often used in the literature.

The following guide to curvilinear coordinates in general provides some of the relations needed later to handle flux coordinates.

The curvilinear coordinates will be labeled $(\theta^1, \theta^2, \theta^3)$. Then

$$\mathbf{e}^i \equiv \nabla \theta^i$$

are called *covariant* basis vectors (not necessarily unit vectors)

$$\mathbf{e}_1 \equiv \nabla \theta^2 \times \nabla \theta^3 / D$$

and its permutations are called *contravariant* basis vectors

$$D \equiv \nabla \theta^1 \cdot \nabla \theta^2 \times \nabla \theta^3$$

is the reciprocal of the *Jacobian*, which is related to the differential volume element in Cartesian coordinates by

$$dx\,dy\,dz = d\theta^1\,d\theta^2\,d\theta^3 / D.$$

Any vector may be expressed in terms of contravariant components a^i,

$$\mathbf{a} = a^i \mathbf{e}_i, \qquad a^i \equiv \mathbf{a} \cdot \mathbf{e}^i,$$

or covariant components a_i, $\mathbf{a} = a_i \mathbf{e}^i$, $a_i \equiv \mathbf{a} \cdot \mathbf{e}_i$. The summation convention is implied on repeated indices. The metric tensors $g_{ij} \equiv \mathbf{e}_i \cdot \mathbf{e}_j$ and $g^{ij} \equiv \mathbf{e}^i \cdot \mathbf{e}^j$ are used to convert between covariant and contravariant components

$$a^i = g^{ij} a_j, \qquad a_i = g_{ij} a^j.$$

Note that

$$g_{ik} g^{kj} = \delta_i^j = \begin{cases} 1 \text{ for } i = j \\ 0 \text{ for } i \neq j \end{cases}$$

and

$$\|g^{ij}\| = |D|^2.$$

Toroidal Instabilities

The following is a basic list of useful identities:

$$\mathbf{a} \cdot \mathbf{b} = a^i b_i = a_i b^i$$

$$(\mathbf{a} \times \mathbf{b})^1 = (a_2 b_3 - a_3 b_2) D \quad \text{and permutations}$$

$$(\mathbf{a} \times \mathbf{b})_1 = (a^2 b^3 - a^3 b^2)/D \quad \text{and permutations}$$

$$\nabla \varphi = \partial \varphi / \partial \theta^i \mathbf{e}^i$$

$$\mathbf{a} \cdot \nabla \varphi = a^i \partial \varphi / \partial \theta^i$$

$$\nabla \cdot (D \mathbf{e}_i) = 0$$

$$\nabla \cdot \mathbf{a} = D \partial (a^i/D)/\partial \theta^i$$

$$\nabla \times \mathbf{e}^i = 0$$

$$(\nabla \times \mathbf{a})^1 = (\partial a_3/\partial \theta^2 - \partial a_2/\partial \theta^3) D \quad \text{and permutations}$$

$$\nabla^2 \varphi = D \frac{\partial}{\partial \theta^i} \left(\frac{g^{ij}}{D} \frac{\partial \varphi}{\partial \theta^j} \right).$$

We now specialize to a *flux coordinate system* (V, θ, ζ) in which V is the volume enclosed by each toroidal flux surface, θ corresponds to an angle the short way around, and ζ corresponds to an angle the long way around the toroid. The toroidal flux surfaces are assumed to be simply nested, at least locally. The coordinate V may be replaced by any surface quantity that uniquely labels each flux surface—such as the poloidal or toroidal flux. The grid lines of constant θ and ζ can be deformed in any way provided they close upon themselves once around the toroid in each direction. It is usually assumed that θ and ζ increase by unity around the toroid in their respective directions.

Since magnetic field lines lie in the flux surfaces ($B \cdot \nabla V = 0$), the magnetic field can be written

$$\mathbf{B} = B^\theta \nabla \zeta \times \nabla V / D + B^\zeta \nabla V \times \nabla \theta / D \tag{7.1.1}$$

where B^θ, B^ζ are the contravariant components of $\mathbf{B}$ and D is the reciprocal of the Jacobian. Then from the condition

$$\nabla \cdot \mathbf{B} = D \left[\frac{\partial}{\partial \theta} \left(\frac{B^\theta}{D} \right) + \frac{\partial}{\partial \zeta} \left(\frac{B^\zeta}{D} \right) \right] = 0 \tag{7.1.2}$$

it follows that the magnetic field may be written in terms of a stream function v

$$\frac{B^\theta}{D} = -\frac{\partial v}{\partial \zeta}, \quad \frac{B^\zeta}{D} = \frac{\partial v}{\partial \theta} \tag{7.1.3}$$

where $v(V,\theta,\zeta)$ may be multivalued in the angles θ and ζ. If v were a quadratic or higher-order function of θ and ζ, the resulting magnetic field would not be single-valued. Therefore, v must be the sum of terms that are linear or periodic in θ and ζ. In particular, from (7.1.1) and (7.1.3) it follows that

$$\mathbf{B} = \nabla V \times \nabla v \qquad (7.1.4)$$

$$v = \theta \dot{\psi}_{\text{tor}}(V) - \zeta \dot{\psi}_{\text{pol}}(V) + \lambda(V,\theta,\zeta) \qquad (7.1.5)$$

where λ is a periodic function in θ and ζ, and ψ_{pol}, ψ_{tor} are the poloidal and toroidal fluxes defined by (4.2.1) and (4.2.2). In order to demonstrate the validity of the coefficients $\dot{\psi}$ in (7.1.5) evaluate

$$\dot{\psi}_{\text{tor}} = \frac{1}{dV} \int_V^{V+dV} d^3x \mathbf{B} \cdot \nabla \zeta = \int_0^1 d\theta \int_0^1 d\zeta B^\zeta/D = \int_0^1 d\theta \int_0^1 d\zeta \frac{\partial v}{\partial \theta}.$$

Question 7.1.1
For a given magnetic field and flux coordinate system, how would the periodic function $\lambda(V,\theta,\zeta)$ be determined? Suppose the surfaces and fluxes are held fixed as λ is varied until the magnetic energy is minimized. What condition does this minimization impose on the current density?

Up to this point, the angles θ, ζ have been arbitrary. Now the coordinate grid will be deformed to eliminate the periodic function λ

$$\theta_{\text{new}} = \theta_{\text{old}} + \lambda/\dot{\psi}_{\text{tor}}(V), \quad \zeta_{\text{new}} = \zeta_{\text{old}}$$

or

$$\theta_{\text{new}} = \theta_{\text{old}}, \quad \zeta_{\text{new}} = \zeta_{\text{old}} - \lambda/\dot{\psi}_{\text{pol}}(V).$$

It follows that

$$v = \theta \dot{\psi}_{\text{tor}}(V) - \zeta \dot{\psi}_{\text{pol}}(V) \qquad (7.1.6)$$

and

$$\mathbf{B} = \dot{\psi}_{\text{tor}}(V) \nabla V \times \nabla \theta + \dot{\psi}_{\text{pol}}(V) \nabla \zeta \times \nabla V. \qquad (7.1.7)$$

If any flux surface is cut open and laid out flat with the angles θ and ζ forming a Cartesian grid, as shown in fig. 7.1, it can be seen that the magnetic field points everywhere in the same direction, so that the field lines appear straight. It is clear from this picture that the q-value is just the ratio of the contravariant components of B

Toroidal Instabilities

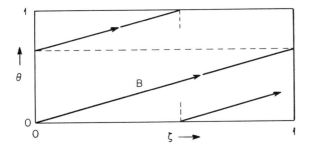

Fig. 7.1
Toroidal flux surface cut open and laid out flat. Field lines appear straight in the appropriate θ, ζ coordinates.

$$q = B^\zeta/B^\theta = \dot\psi_{\text{tor}}/\dot\psi_{\text{pol}} = d\psi_{\text{tor}}/d\psi_{\text{pol}} \tag{7.1.8}$$

as derived before in section 4.3. These flux coordinates are discussed in more detail by Solov'ev and Shafranov (1970) among others.

Two more conditions must be satisfied in order to construct Hamada coordinates—the most frequently used flux coordinates in the literature. First, the Jacobian must be normalized to unity everywhere

$$D = \nabla V \cdot \nabla\theta \times \nabla\zeta = 1 \tag{7.1.9}$$

so that $dx\,dy\,dz = dV\,d\theta\,d\zeta$. A proof that the coordinate system can always be deformed to satisfy (7.1.9), as long as $p'(V) \neq 0$, is adequately discussed by Greene and Johnson (1962) and by Solov'ev and Shafranov (1970).

Up to this point, flux coordinates could be used for any magnetic field with simply nested flux surfaces. The final condition for Hamada coordinates is that the magnetic field must correspond to an MHD equilibrium ($\mathbf{J} \times \mathbf{B} = \nabla p$). Then current density lines as well as magnetic field lines appear as straight lines on the θ, ζ plane. To show this, use the fact that no current passes through flux surfaces in an MHD equilibrium to write

$$\mathbf{J} = J^\theta \nabla\zeta \times \nabla V + J^\zeta \nabla V \times \nabla\theta \tag{7.1.10}$$

where, at this point, J^θ and J^ζ are functions of V, θ, ζ. With $D = 1$, the magnetic field has the form

$$\mathbf{B} = B^\theta(V)\nabla\zeta \times \nabla V + B^\zeta(V)\nabla V \times \nabla\theta$$

where $B^\theta(V) = \dot\psi_{\text{pol}}(V)$ and $B^\zeta(V) = \dot\psi_{\text{tor}}(V)$. Then the equilibrium condition $\mathbf{J} \times \mathbf{B} = \nabla p$ implies

$$J^\theta B^\zeta(V) - J^\zeta B^\theta(V) = p'(V). \tag{7.1.11}$$

Take θ and ζ derivatives of this expression and use

$$\mathbf{V} \cdot \mathbf{J} = \partial J^\theta / \partial \theta + \partial J^\zeta / \partial \zeta = 0 \tag{7.1.12}$$

to show that $\mathbf{B} \cdot \mathbf{V} J^\theta = \mathbf{B} \cdot \mathbf{V} J^\zeta = 0$. It follows that the contravariant components J^θ and J^ζ are surface quantities and therefore the current density lines appear as straight lines on the θ, ζ plane. Also, it can be shown that

$$J^\theta = \dot{I}_{\text{pol}}(V) \quad \text{and} \quad J^\zeta = \dot{I}_{\text{tor}}(V) \tag{7.1.13}$$

where $I_{\text{pol}}(V)$ and $I_{\text{tor}}(V)$ are the total poloidal and toroidal currents associated with each flux surface defined by (4.2.3) and (4.2.4).

In summary then, Hamada coordinates are flux coordinates applicable to MHD equilibria in which both the B-field and J-field lines appear straight and the Jacobian is normalized to unity everywhere. In general, Hamada coordinates are not orthogonal and they generally depend upon the flux profiles ($\psi_{\text{pol}}(V)$ and $\psi_{\text{tor}}(V)$) as well as the geometry of the flux surfaces. They were first used by Hamada in a paper whose English translation appeared in 1962.

Question 7.1.2
How would you go about constructing an explicit representation for Hamada coordinates $V(R, y, \varphi)$, $\theta(R, y, \varphi)$, and $\zeta(R, y, \varphi)$, given an axisymmetric toroidal equilibrium specified by $\psi_{\text{pol}} = \psi(R, y)$, $p = p(\psi)$, and $RB_\varphi = I(\psi)$, as in chapter 4—such as the Shafranov equilibrium (4.4.15). Don't worry about actually carrying out the integrations or inverting the functions; assume that powerful analytic or computational methods are available to carry out your plan.

7.2 MERCIER STABILITY CRITERION

The Mercier stability criterion is a direct generalization of the Suydam criterion (6.5.16) to toroidal plasmas with any shape, any aspect ratio, and any plasma beta. It is a criterion for the stability of modes localized arbitrarily close to any mode rational surface within the plasma. As such, it is a *necessary* condition for linear ideal MHD stability. When the Mercier criterion indicates instability, there will certainly be at least a localized instability and possibly a large scale instability as well. When the Mercier criterion indicates stability, a large scale instability may still remain. A *sufficient* condition for stability would be needed

Toroidal Instabilities

to ensure the stability of all MHD modes. Of course, the Mercier criterion says nothing about how the instabilities will evolve under more realistic nonideal and nonlinear conditions.

A detailed derivation of the Mercier criterion, closely following Greene and Johnson (1962), will be given here. For an alternative derivation see Mercier and Luc (1974) and for additional physical insight see Greene and Johnson (1968). The techniques used in this derivation are widely used in the analytic literature on toroidal instabilities. The criterion itself is now used as a routine test of computed equilibria.

The derivation begins with the energy principle (5.4.7)

$$\delta W = \tfrac{1}{2}\int dV\,d\theta\,d\zeta\,\{|\mathbf{B}^1 + \boldsymbol{\xi}\cdot\nabla V\mathbf{J}\times\nabla V/|\nabla V|^2|^2 \\ + \Gamma p|\nabla\cdot\boldsymbol{\xi}|^2 + K|\boldsymbol{\xi}\cdot\nabla V|^2\} \quad (7.2.1)$$

where

$$\mathbf{B}^1 = \nabla\times(\boldsymbol{\xi}\times\mathbf{B}) \quad (7.2.2)$$

is the perturbed magnetic field (5.2.5) and

$$K \equiv 2\mathbf{J}\times\nabla V\cdot(\mathbf{B}\cdot\nabla)\nabla V/|\nabla V|^4 \quad (7.2.3)$$

is the only term in δW which can be negative and therefore destabilizing.

Everything will be written in terms of Hamada coordinates (V,θ,ζ) which were described in the last section. Of particular importance are the expressions for the equilibrium quantities

$$\mathbf{B} = B^\theta(V)\mathbf{e}_\theta + B^\zeta(V)\mathbf{e}_\zeta \quad (7.2.4)$$

$$\mathbf{J} = J^\theta(V)\mathbf{e}_\theta + J^\zeta(V)\mathbf{e}_\zeta \quad (7.2.5)$$

$$p'(V) = J^\theta B^\zeta - J^\zeta B^\theta. \quad (7.2.6)$$

Of the three contravariant basis vectors

$$\mathbf{e}_V = \nabla\theta\times\nabla\zeta,\quad \mathbf{e}_\theta = \nabla\zeta\times\nabla V,\quad \mathbf{e}_\zeta = \nabla V\times\nabla\theta \quad (7.2.7)$$

two can be expressed in terms of the equilibrium fields

$$\mathbf{e}_\theta = (B^\zeta\mathbf{J} - J^\zeta\mathbf{B})/p'(V) \\ \mathbf{e}_\zeta = (-B^\theta\mathbf{J} + J^\theta\mathbf{B})/p'(V) \quad (7.2.8)$$

provided $\mathbf{B}$ and $\mathbf{J}$ are not both in the same direction.

Decompose the displacement vector in the following way

$$\boldsymbol{\xi} = \xi^V\mathbf{e}_V + \xi^\theta\mathbf{e}_\theta + \xi^\zeta\mathbf{e}_\zeta \quad (7.2.9)$$

$$\mu \equiv B^\zeta\xi^\theta - B^\theta\xi^\zeta \quad (7.2.10)$$

$$\eta \equiv -J^\zeta \xi^\theta + J^\theta \xi^\zeta \qquad (7.2.11)$$

noting that $\mu = (\xi \times \mathbf{B})_V$ and $\eta = (\mathbf{J} \times \xi)_V$. The contravariant components of the displacement vector are

$$\begin{aligned}\xi^V &= \xi \cdot \nabla V \\ \xi^\theta &= (B^\theta \eta + J^\theta \mu)/p'(V) \\ \xi^\zeta &= (B^\zeta \eta + J^\zeta \mu)/p'(V).\end{aligned} \qquad (7.2.12)$$

It follows that the components of the electric field are

$$\begin{aligned}(\xi \times \mathbf{B})_V &= \mu \\ (\xi \times \mathbf{B})_\theta &= -\xi^V B^\zeta \\ (\xi \times \mathbf{B})_\zeta &= \xi^V B^\theta\end{aligned} \qquad (7.2.13)$$

and the perturbed magnetic field can be shown to be

$$\mathbf{B}^1 = \left[B^\theta \frac{\partial}{\partial \theta} \xi^V + B^\zeta \frac{\partial}{\partial \zeta} \xi^V \right] \mathbf{e}_V + \left[\frac{\partial}{\partial \zeta} \mu - \frac{\partial}{\partial V}(\xi^V B^\theta) \right] \mathbf{e}_\theta \\ + \left[-\frac{\partial}{\partial V}(\xi^V B^\zeta) - \frac{\partial}{\partial \theta} \mu \right] \mathbf{e}_\zeta. \qquad (7.2.14)$$

Finally, the divergence of the displacement vector, corresponding to compression or expansion, has the form

$$\nabla \cdot \xi = \frac{\partial}{\partial V} \xi^V + \frac{\partial}{\partial \theta} \xi^\theta + \frac{\partial}{\partial \zeta} \xi^\zeta. \qquad (7.2.15)$$

Now choose any mode numbers m and n, such that there is a corresponding mode rational surface within the plasma with

$$q_0 = B_0^\zeta / B_0^\theta = m/n. \qquad (7.2.16)$$

All equilibrium quantities at this surface will be designated with the subscript 0.

The central assumption in this derivation is that the perturbation is localized to within an arbitrarily small neighborhood $V_0 + \varepsilon > V > V_0 - \varepsilon$ of this mode rational surface. A new variable will be defined on the small scale length of the instability

$$x \equiv (V - V_0)/\varepsilon \qquad (7.2.17)$$

defined so that

$$|\xi| \to 0 \quad \text{as} \quad |x| \to 1. \qquad (7.2.18)$$

Toroidal Instabilities

The derivation will proceed by a systematic expansion of the energy principle in powers of ε. All equilibrium variables will be expanded as a power series around the mode rational surface

$$B^\theta = B_0^\theta + \varepsilon x \dot{B}_0^\theta + \cdots \text{ etc.} \tag{7.2.19}$$

where B_0^θ and $\dot{B}_0^\theta$ are numerical constants, independent of x, θ, and ζ. This procedure will lead to an analytic minimization of δW, order by order in ε. All perturbed quantities will also be expanded using the notation

$$\xi^V = \xi_0^V(x,\theta,\zeta) + \varepsilon \xi_1^V(x,\theta,\zeta) + \cdots \text{ etc.} \tag{7.2.20}$$

Using $(\partial/\partial V) = (1/\varepsilon)(\partial/\partial x)$ and $dV d\theta d\zeta = \varepsilon dx d\theta d\zeta$, we find that the lowest order contributions to δW come from the ε^{-1} orders of the terms $\mathbf{V}\cdot\boldsymbol{\xi}$ and $\mathbf{B}^1$,

$$(\mathbf{V}\cdot\boldsymbol{\xi})_{-1} = \partial \xi_0^V / \partial x$$

$$(\mathbf{B}^1)_{-1} = -\left[B_0^\theta \frac{\partial}{\partial x}\xi_0^V \mathbf{e}_{\theta 0} + B_0^\zeta \frac{\partial}{\partial x}\xi_0^V \mathbf{e}_{\zeta 0}\right],$$

while the $\boldsymbol{\xi}\cdot\mathbf{V}V = \xi_0^V$ term is only zero order in ε. It follows that the lowest order of δW is the ε^{-1} order

$$\delta W_{-1} = \tfrac{1}{2}\int dx d\theta d\zeta \{|\mathbf{B}_{-1}^1|^2 + \Gamma p_0 |(\mathbf{V}\cdot\boldsymbol{\xi})_{-1}|^2\}. \tag{7.2.21}$$

These positive terms in δW_{-1} can be made to vanish if

$$\partial \xi_0^V / \partial x = 0.$$

Since this condition must be forced to apply over the full range $-1 < x < +1$, it follows from the boundary condition (7.2.18) that

$$\xi_0^V = 0. \tag{7.2.22}$$

Physically, the leading order flow and its large gradient perpendicular to the flux surfaces must be suppressed in order to avoid large compression which rapidly increases both thermodynamic and magnetic energy.

Question 7.2.1
Why couldn't the compression be avoided by having huge flows parallel to the flux surfaces such that $\mu, \eta \sim \mathcal{O}(\varepsilon^{-1})$?

Going to the next order in ε, it is found that the zero order parts of $\mathbf{B}^1$ and $\mathbf{V}\cdot\boldsymbol{\xi}$ have the form

Toroidal Instabilities

$$(\mathbf{B}^1)_0 = \left[\frac{\partial}{\partial \zeta}\mu_0 - B_0^\theta \frac{\partial}{\partial x}\xi_1^V\right]\mathbf{e}_{\theta 0} + \left[-B_0^\zeta \frac{\partial}{\partial x}\xi_1^V - \frac{\partial}{\partial \theta}\mu_0\right]\mathbf{e}_{\zeta 0} \quad (7.2.23)$$

$$(\nabla \cdot \boldsymbol{\xi})_0 = \frac{\partial}{\partial x}\xi_1^V + \frac{\partial}{\partial \theta}\xi_0^\theta + \frac{\partial}{\partial \zeta}\xi_0^\zeta$$

$$= \frac{\partial}{\partial x}\xi_1^V + (\mathbf{J}_0 \cdot \nabla \mu_0 + \mathbf{B}_0 \cdot \nabla \eta_0)/p_0', \quad (7.2.24)$$

which can be shown by substitution, while the zero order part of $(\boldsymbol{\xi} \cdot \nabla V)_0 = \xi_0^V$ vanishes. These fields combine to make a first order contribution to δW, similar in form to (7.2.21), which can be made to vanish if $B_0^{\theta 1} = B_0^{\zeta 1} = 0$ and $(\nabla \cdot \boldsymbol{\xi})_0 = 0$. The vanishing of these magnetic field components yields

$$\mathbf{B}_0 \cdot \nabla \mu_0 = 0 \quad (7.2.25)$$

and

$$\frac{\partial}{\partial x}\mathbf{B}_0 \cdot \nabla \xi_1^V = 0.$$

Applying this second condition over the range $-1 < x < +1$ yields

$$\mathbf{B}_0 \cdot \nabla \xi_1^V = 0. \quad (7.2.26)$$

Equations (7.2.25) and (7.2.26) imply that the leading order radial perturbation ξ_1^V and radial electric field μ_0 are uniform along each magnetic field line $\mathbf{B}_0$ in the mode rational surface. But they may vary arbitrarily perpendicular to the field lines within the surface as well as perpendicular to the flux surfaces. In effect, different field lines in the mode rational surface are independent of each other. Because of this, it is useful to define a coordinate which is constant along each field line $\mathbf{B}_0$ within the mode rational surface

$$u \equiv m\theta - n\zeta \quad (7.2.27)$$

so that, from (7.2.16)

$$\mathbf{B}_0 \cdot \nabla u = mB_0^\theta - nB_0^\zeta = 0. \quad (7.2.28)$$

This coordinate u is then extended so that (7.2.28) applies to all the flux surfaces, even though $\mathbf{B}_0$ refers to only the magnetic field in the mode rational surface. Using this new coordinate, (7.2.25) and (7.2.26) imply

$$\mu_0 = \mu_0(x, u) \quad (7.2.29)$$

Toroidal Instabilities   135

$$\xi_1^V = \xi_1^V(x, u). \tag{7.2.30}$$

Then, setting $(\mathbf{V} \cdot \boldsymbol{\xi})_0 = 0$, from (7.2.24) we find

$$\eta_0 = \eta_0(x, u). \tag{7.2.31}$$

Again using (7.2.24) and

$$\mathbf{J}_0 \cdot \nabla \mu_0(x, u) = (J_0^\theta m - J_0^\zeta n) \partial \mu_0 / \partial u = m p_0' / B_0^\zeta$$

we find

$$\frac{\partial}{\partial x} \xi_1^V + \frac{m}{B_0^\zeta} \frac{\partial \mu_0}{\partial u} = 0. \tag{7.2.32}$$

The Mercier criterion results from minimizing δW to the next higher order in ε. To do this, we resolve the perturbed magnetic field $\mathbf{B}^1$ along the orthogonal directions ∇V_0, $\mathbf{B}_0$, and $\mathbf{B}_0 \times \nabla V_0$. It is shown by Greene and Johnson (1962) and by Mercier and Luc (1974) that the components of the flow parallel to the mode rational surface can be adjusted so that there is no compression to this order

$$(\mathbf{V} \cdot \boldsymbol{\xi})_1 = 0 \tag{7.2.33}$$

and so that the component parallel to $\mathbf{B}_0$ of the first term in δW (7.2.1) vanishes

$$\mathbf{B}_0 \cdot (\mathbf{B}_1^1 + \xi_1^V \mathbf{J}_0 \times \nabla V_0 / |\nabla V_0|^2) = 0 \tag{7.2.34}$$

without affecting the minimization by the other components of the flow. All we have left to minimize is

$$\delta W_3 = \int dx \, d\theta \, d\zeta \left\{ \left| \left(\mathbf{B}_1^1 + \xi_1^V \frac{\mathbf{J}_0 \times \nabla V_0}{|\nabla V_0|^2} \right) \cdot \frac{\mathbf{B}_0 \times \nabla V_0}{|\mathbf{B}_0 \times \nabla V_0|} \right|^2 \right.$$
$$\left. + K_0 (\xi_1^V)^2 \right\} \tag{7.2.35}$$

where K_0 is (7.2.3) evaluated at the mode rational surface.

To find the other two components of the perturbed magnetic field, use (7.2.14) and (7.2.8) to show

$$\mathbf{B}^1 = \mathbf{B} \cdot \nabla \xi^V \mathbf{e}_V$$
$$+ \frac{1}{p'} [\mathbf{B} \cdot \nabla \mu + (B^\theta \dot{B}^\zeta - B^\zeta \dot{B}^\theta) \xi^V] \mathbf{J}$$
$$+ \frac{1}{p'} \left[-\mathbf{J} \cdot \nabla \mu + J^\zeta \frac{\partial}{\partial V} (\xi^V B^\theta) + J^\theta \frac{\partial}{\partial V} (\xi^V B^\zeta) \right] \mathbf{B}.$$

Evaluating the required components of $\mathbf{B}^1$ to first order in ε, we find

$$(\mathbf{B} \cdot \nabla \xi^V)_1 = \mathbf{B}_0 \cdot \nabla \xi_1^V(x, u) = 0$$

and

$$\mathbf{B}_1^1 \cdot \frac{\mathbf{B}_0 \times \nabla V_0}{|\mathbf{B}_0 \times \nabla V_0|} = \mathbf{B}_0 \cdot \nabla \mu_1 + \mathbf{B}_1 \cdot \nabla \mu_0$$

$$+ (\dot{B}_0^\theta \dot{B}_0^\zeta - \dot{B}_0^\zeta \dot{B}_0^\theta) \xi_1^V \quad (7.2.36)$$

where $\dot{B} = \dot{B}(x)$. Using (7.2.29), (7.2.32), and (7.2.16), it can be shown that

$$\mathbf{B}_1 \cdot \nabla \mu_0 = x \left(\dot{B}_0^\theta \frac{\partial}{\partial \theta} + \dot{B}_0^\zeta \frac{\partial}{\partial \zeta} \right) \mu_0(x, u) = x S_0 \partial \xi_1^V / \partial x$$

where

$$S_0 \equiv B_0^\theta \dot{B}_0^\zeta - B_0^\zeta \dot{B}_0^\theta \quad (7.2.37)$$

is known as *global shear*. After all this derivation, the third order contribution to δW reduces to

$$\delta W_3 = \int dx \, d\theta \, d\zeta \left\{ \frac{1}{N} |\mathbf{B}_0 \cdot \nabla \mu_1 + G|^2 + K_0 (\xi_1^V)^2 \right\} \quad (7.2.38)$$

where

$$G \equiv S_0 \frac{\partial}{\partial x}(x \xi_1^V) + \xi_1^V \mathbf{J}_0 \cdot \mathbf{B}_0 / |\nabla V_0|^2$$

and

$$N \equiv B_0^2 / |\nabla V_0|^2.$$

This expression for δW_3 will first be minimized with respect to the term

$$\sigma \equiv \mathbf{B}_0 \cdot \nabla \mu_1 \quad (7.2.39)$$

and then with respect to ξ_1^V. Equation (7.2.39) is known as a *magnetic differential equation* for μ_1, given σ and $\mathbf{B}_0$. If μ_1 were known at any point along B_0, its value at any other point along the same field line would be

$$\mu_1 = \int_0^l \frac{dl}{|B_0|} \sigma + \mu_1(l = 0) \quad (7.2.40)$$

Toroidal Instabilities

where l is the arclength along the field line. For μ_1 to be single valued around any closed field line $\mathbf{B}_0$, σ must satisfy the condition

$$\oint \frac{dl}{|\mathbf{B}_0|} \sigma = 0. \tag{7.2.41}$$

Therefore, the minimization of δW_3 with respect to σ must be carried out using (7.2.41) as a constraint. This constraint, which is equivalent to

$$\int d\theta \, d\zeta \, \delta(u - u_0)\sigma(V, \theta, \zeta) = 0$$

can be implemented in the minimization by using the variational form

$$\delta[\delta W_3 + \int dV \, d\theta \, d\zeta \, \lambda(V, u) \, \sigma(V, \theta, \zeta)] = 0 \tag{7.2.42}$$

where λ is a Lagrange multiplier to be determined. This variation leads to

$$\frac{2}{N}(\sigma + G) + \lambda = 0$$

where G and N are defined after (7.2.38). Then applying condition (7.2.41) yields

$$\lambda = -2\langle G \rangle / \langle N \rangle$$

where the symbol $\langle \cdots \rangle \equiv \oint dl/|\mathbf{B}_0| \cdots / \oint dl/|\mathbf{B}_0|$ stands for the average over a closed field line. After this minimization, δW_3 assumes the form

$$\delta W_3 = \int dx \, d\theta \, d\zeta \left\{ \left[S_0 + \frac{\partial}{\partial x}(x\xi_1^V) + \xi_1^V \langle \mathbf{J}_0 \cdot \mathbf{B}_0 / |\nabla V_0|^2 \rangle \right]^2 \frac{1}{\langle N \rangle} \right.$$
$$\left. + K_0(\xi_1^V)^2 \right\} \tag{7.2.43}$$

where $\langle \xi_1^V \rangle = \xi_1^V$ because $\xi_1^V = \xi_1^V(x, u)$.

Now δW_3 has been reduced to the point where all the coefficients are numerical constants independent of x. The final minimization can be accomplished by setting

$$\xi_1^V(x, u) = \xi(x)f(u) \tag{7.2.44}$$

where $f(u)$ represents an arbitrary variation of the perturbation within the flux surface perpendicular to the mode rational magnetic field $\mathbf{B}_0$. Now a test function can be constructed for $\xi(x)$ which is identical to the test function used to derive the Suydam criterion (6.5.16). In particular, we can use the inequality

$$\int dx \left[\frac{d}{dx}(x\xi)\right]^2 \geq \tfrac{1}{4}\int dx\,\xi^2, \tag{7.2.45}$$

where the equality is approached arbitrarily closely by the minimizing function (see Greene and Johnson (1962), Appendix III). The Mercier stability criterion then follows from (7.2.43)–(7.2.45)

$$[\tfrac{1}{2}S_0 + \langle \mathbf{J}_0 \cdot \mathbf{B}_0/|\nabla V_0|^2\rangle]^2 + \langle K_0\rangle\langle B_0^2/|\nabla V_0|^2\rangle \geq 0 \tag{7.2.46}$$

where the global shear S_0 and the driving term K_0 are defined by (7.2.37) and (7.2.3) evaluated at the mode rational surface.

A more commonly used expression for the Mercier criterion was derived by Solov'ev (1968) who showed that $\langle K_0\rangle$ can be rearranged

$$\langle K_0\rangle = -\langle J_0^2/|\nabla V_0|^2\rangle - \Omega_0 \tag{7.2.47}$$

where

$$\Omega_0 \equiv J_0^\theta \dot{B}_0^\zeta - J_0^\zeta \dot{B}_0^\theta \tag{7.2.48}$$

is known as a measure of the magnetic well. The Mercier stability criterion can then be written

$$\tfrac{1}{4}S^2 + S\langle \mathbf{J}\cdot\mathbf{B}/|\nabla V|^2\rangle - \Omega\langle B^2/|\nabla V|^2\rangle$$
$$-(\langle J^2/|\nabla V|^2\rangle\langle B^2/|\nabla V|^2\rangle - \langle \mathbf{J}\cdot\mathbf{B}/|\nabla V|^2\rangle^2) > 0 \tag{7.2.49}$$

This necessary stability criterion can be evaluated at any rational surface within the plasma. Reviewing the symbols used in this expression,

$$S \equiv B^\theta \dot{B}^\zeta - B^\zeta \dot{B}^\theta = (B^\theta)^2 \frac{d}{dV}q(V)$$

measures the shear of the magnetic field, where $B^\theta = \dot{\psi}_{\mathrm{pol}}(V)$, $B^\zeta = \dot{\psi}_{\mathrm{tor}}(V)$,

$$\langle \cdots \rangle \equiv \oint \frac{dl}{|B|}\cdots \Big/ \oint \frac{dl}{|B|}$$

is a line average along the magnetic field at the rational surface, V is the volume (or any labeling) of the flux surfaces, $\mathbf{J}$ and $\mathbf{B}$ are the equilibrium current density and magnetic field at the rational surface, and $\Omega \equiv J^\theta \dot{B}^\zeta - J^\zeta \dot{B}^\theta$ is a measure of the magnetic well, where $J^\theta = \dot{I}_{\mathrm{pol}}(V)$ and $J^\zeta = \dot{I}_{\mathrm{tor}}(V)$.

I am indebted to Günther Spies for a concise introduction to flux coordinates and to John Johnson and David Nelson for advice on the Mercier stability derivation.

7.3 APPLICATIONS OF THE MERCIER CRITERION

The formal expressions (7.2.49) and (7.2.46) just derived for the Mercier criterion apply to any magnetically confined plasma with nested flux surfaces and shear. However, when people speak about the Mercier criterion they often refer to one of the simplified expressions representing an approximation for particular equilibria. The most commonly used approximate expression for the Mercier criterion is the expression by Shafranov and Yurchenko (1968) for axisymmetric tokamaks with large aspect ratio, low beta and circular cross section

$$\frac{1}{4}\left(\frac{q'(r)}{q(r)}\right)^2 + \frac{2\mu p'(r)}{rB_\varphi^2}(1 - q^2) > 0 \qquad (7.3.1)$$

where r is the minor radius of what are assumed to be circular flux surfaces and B_φ is the toroidal component of the magnetic field evaluated at the magnetic axis. The derivation of (7.3.1) includes the fact that the inner flux surfaces are shifted outward along the major radius relative to the outer surfaces.

The first term in (7.3.1) represents the stabilizing effect of shear—just as in the Suydam criterion. When the pressure is decreasing away from the magnetic axis, the second term is stabilizing if $q^2 > 1$ and destabilizing if $q^2 < 1$—just as in the Kruskal-Shafranov condition. This dependence on q^2 distinguishes the toroidal Mercier criterion from the straight cylinder Suydam criterion. According to Shafranov and Yurchenko (1968), this stabilizing q^2 term results from the combination of three effects.

1. The average toroidal curvature is stabilizing. To see this, we can approximate the curvature of a field line by

$$\kappa \equiv \hat{\mathbf{B}} \cdot \nabla \hat{\mathbf{B}} \simeq \frac{1}{|B|}\left(\frac{B_\theta}{r}\frac{\partial}{\partial\theta} + \frac{B_\varphi}{R}\frac{\partial}{\partial\varphi}\right)\frac{B_\theta \hat{\theta} + B_\varphi \hat{\varphi}}{|B|}$$

$$\simeq -\frac{1}{|B|^2}\left(B_\theta^2 \frac{\hat{\mathbf{r}}}{r} + B_\varphi^2 \frac{\hat{\mathbf{R}}}{R}\right)$$

where θ is the angle around the circular flux surface of radius r. The first term is the poloidal curvature, with radius of curvature r, and the second term is the toroidal curvature, with radius of curvature R. The effect of curvature on the potential energy can be seen in the last term of (5.4.9). Since the radial component of the perturbation is essentially uniform along the field lines near the mode rational surface, the poloidal part of the curvature makes a destabilizing contribution to

(5.4.9). But the toroidal curvature makes a stabilizing (negative) contribution at the inner edge of the torus where it is largest, and a destabilizing contribution at the outer edge where it is smaller, so that the average effect is stabilizing.

2. There is a magnetic well effect because the inner flux surfaces are shifted outward along the major radius, putting them in a lower toroidal magnetic field. As a result, the radial perturbation tends to push the plasma into a stronger magnetic field.

3. The poloidal variation of the pitch of the field lines has a slightly stabilizing effect. It should be kept in mind that these concepts, such as magnetic well and curvature stabilization, are merely interpretations based on certain rearrangements of the mathematical terms. They cannot be used alone to make predictions.

By comparing (7.3.1) with computer evaluations of the formal expression for the Mercier criterion (7.2.49), it is found that (7.3.1) is not very accurate away from the magnetic axis for aspect ratios less than 15 or 20, while tokamaks are typically built with aspect ratios between 3 and 5. This is not a serious problem, however, because the Mercier criterion usually predicts instability first at the magnetic axis, as the equilibrium parameters are varied, before it predicts instability away from the magnetic axis.

Question 7.3.1
Is it a coincidence that the Mercier criterion, which was derived for radially localized instabilities, gives essentially the same stability criterion ($q_{axis} > 1$) as for the internal $m=1$ kink mode?

Question 7.3.2
The $m=2$, $n=1$ instability is localized near the $q=2$ rational surface, which is generally in a high shear region away from the magnetic axis. Does the Mercier criterion suggest that this mode is always stable?

Question 7.3.3
Given the fact that the Mercier criterion is clearly different from the Suydam criterion, how does such a fine scale instability know that it is living on a torus and not on a straight cylinder—even if the torus has arbitrarily large aspect ratio?

Another simple evaluation of the Mercier criterion, by Lortz and Nührenberg (1973), is based on a series expansion in the neighborhood

Toroidal Instabilities

of the magnetic axis for axisymmetric equilibria with arbitrary cross section and arbitrary poloidal beta. In addition, the sufficient stability criterion by Lortz (1973) is cast in the same form so that it can be easily compared with the necessary Mercier criterion. Presumably, the necessary and sufficient ideal MHD stability criterion in the neighborhood of the magnetic axis must lie somewhere between these two criteria. The combined criteria are

$$\frac{1}{q^2} < \frac{6}{1+e^2} - \frac{4}{e(e+\delta)} + Q\left[\frac{4}{e(e+\delta)} - 2\right] + \frac{e^2-1}{e^2+1}d \qquad (7.3.2)$$

where $\delta = 0$ for the Lortz sufficient criterion and $\delta = 1$ for the Mercier necessary criterion. Here, the q-value is the standard form (7.1.8), $e = l_y/l_R$ is the vertical elongation of the flux surfaces in the neighborhood of the magnetic axis, and

$$Q \equiv J^\theta B^\zeta/(J^\zeta B^\theta) = \dot{I}_{pol}\dot{\psi}_{tor}/\dot{I}_{tor}\dot{\psi}_{pol}$$

is a local measure of $1 - \beta_{pol}$. The symbol d is related to the shift and triangularity of the flux surfaces near the magnetic axis. In order to calculate d, it is necessary to have an expansion for the shape of the flux surfaces in the neighborhood of the magnetic axis

$$V = 2\pi^2(R_a - 2S(R - R_a) + \cdots)[e(R - R_a)^2 + y^2/e + 2\Delta(R - R_a)y^2 + \cdots] \qquad (7.3.3)$$

where V is the volume of the flux surfaces and R_a is the major radius to the magnetic axis. The parameter S, which is a measure of the shift of the flux surfaces relative to the magnetic axis, and the parameter Δ, which is a measure of the triangularity of the flux surfaces, are related to each other through the parameters d and Q by

$$S = \tfrac{1}{6}d - \tfrac{1}{2} \qquad (7.3.4)$$

$$\Delta = \frac{1}{2e} - Q\left(e + \frac{1}{e}\right) + \frac{d}{2}\left(e + \frac{1}{3e}\right). \qquad (7.3.5)$$

Question 7.3.4 (Lortz and Nührenberg (1973))
As the elongation, triangularity, and poloidal beta are varied, what conditions are most favorable for stability near the magnetic axis? Under what conditions do the necessary and the sufficient criteria predict essentially the same result? What is the lowest q-value that can be achieved for a stable plasma?

7.4 LARGE-SCALE INSTABILITIES IN AXISYMMETRIC TORI

There are many papers in the literature that use analytic approximations to determine the marginal stability conditions in toroidal geometry. In the pioneering work of Ware and Haas (1966) and Ware (1964, 1971), an expansion of the energy principle up to sixth order in the reciprocal of the aspect ratio (a/R) is used to determine the stability condition for fixed boundary instabilities in a low beta circular torus; the results are essentially the same as the Mercier criterion. Frieman et al. (1973) used the methods of Ware and Haas to investigate a free-boundary circular toroidal plasma with very low shear; they find that toroidicity has only a small effect on the instabilities, even when the wall is close to the plasma. Bussac et al. (1975) were the first to show analytically that the internal $m=1$, $n=1$ kink instability is stabilized by toroidal effects when β_{pol} is small. The involved algebra and methods of approximation which go into these calculations is beyond the scope of this book.

Instead of presenting more analytic results here, it would be useful to review some of the computational results on the growth rate and spatial structure of large scale instabilities in toroidal geometry. Ideal MHD instability results will be given for moderately low beta, nearly circular tokamaks in this section, and for high beta and noncircular tokamaks in the next chapter.

Essentially, there are two computational methods used for linear stability analysis—the initial value method and the eigenvalue method. For the initial value method, an arbitrary initial perturbation is given and the linearized equations are integrated forward in time until the fastest growing instability dominates over all other motion. If the equilibrium is stable, the perturbation oscillates in time at a frequency which is usually dominated by the Alfvén transit time. The method is implemented by specifying the equilibrium and the perturbed variables on a grid and writing the MHD equations in finite difference form. The method was developed for linear toroidal instabilities by Wesson and Sykes (1974) at Culham and by Schneider and Bateman (1974) at Garching. As we shall see in chapters 9 and 10, the initial value method is used extensively at many laboratories to study the nonlinear and nonideal evolution of MHD instabilities.

For the eigenvalue method, the perturbation is written as a series of basis functions (such as Fourier harmonics) or finite elements (which are functions with adjustable parameters, each used over a small

Toroidal Instabilities 143

element of space) with exponential time dependence. The MHD equations are then turned into a huge matrix equation which is solved for the eigenfunctions and eigenvalues (growth rates and oscillation frequencies). The degree to which this finite set of eigenvalues approximates the full spectrum for the MHD equations can be tested by increasing the number or by changing the choice of basis functions. This method has been heavily developed at Princeton (1976) and at Lausanne (1976).

Both methods work best when neither the eigenfunctions nor the equilibrium have fine scale structure. This rules out the Mercier-Suydam type of perturbations. Both methods have trouble with large aspect ratio, low beta configurations, where growth rates are very small compared to the oscillation frequencies of magnetosonic, Alfvén and sound waves. Under these conditions, the initial value codes take a long time to converge and the eigenvalue codes suffer accuracy problems because of the wide spread in the magnitude of the eigenvalues. For these reasons, there is very little overlap between analytic results and computed results.

Question 7.4.1
Only a finite number of eigenvalues and eigenfunctions can be computed. But there are an infinite number of eigenvalues and eigenfunctions associated with waves resonating at different flux surfaces. How do the computed eigenvalues approximate these continua?

Probably the clearest effect of toroidicity is the fact that the unstable eigenfunctions exhibit a mix of poloidal harmonics. The displacement or velocity flow field is shifted out along the major radius and either a dead space or additional pairs of auxiliary vortices are found near the inner edge of the torus. This is especially true for fixed boundary instabilities at low aspect ratio (fig. 7.2) or high poloidal beta. However, there is a tendency for the instabilities to be stronger near the outer edge of the torus and for additional harmonic structure to appear near the inner edge even for free-boundary instabilities where there is no wall forcing the vortex structure to remain within the plasma. Evidently, the strongest part of the instability shifts to where curvature and parallel current driving terms are largest.

The effect of poloidal harmonic mixing is most striking in the poloidal part of the perturbed magnetic field. At low aspect ratios, the dominant part of the perturbed poloidal magnetic field is one harmonic higher than the dominant part of the displacement field. For

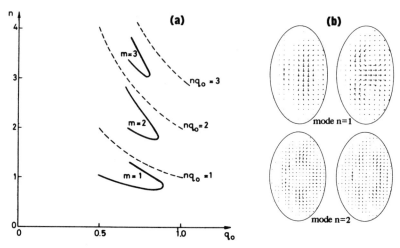

Fig. 7.2
Stability diagram for fixed-boundary modes with eigenfunctions shown for $q_0 = q_{\text{axis}} = 0.81$ in a mildly elongated ($b/a = 1.6$), low aspect ratio ($R/a = 2.5$) tokamak with rounded profile ($p'(\psi) \sim \psi$, $II'(\psi) = 0$). At integer values the continuous variable n is the toroidal mode number. From J. A. Wesson and A. Sykes, IAEA Tokyo Conference, *1*, 449 (1974).

example, if two velocity vortex cells dominate the motion of the plasma, there will be a large positive and a large negative peak in the perturbed pressure and the perturbed toroidal magnetic field, but there will be four vortex cells of comparable magnitude for the perturbed poloidal field. These correspond to four filaments of toroidal perturbed current.

Question 7.4.2
Consider one harmonic of the perturbation displacement

$$\xi = \xi(r) \cos(m\theta - n\varphi)$$

where θ is the poloidal angle around the magnetic axis. How large is the $m \pm 1$ harmonic of $\mathbf{B}_{\text{pol}}^1$ relative to the mth harmonic as a function of the inverse aspect ratio $\varepsilon \equiv r/R$?

As a result of this harmonic mixing, it is no longer possible to make a unique identification of the instabilities on the basis of poloidal mode number. There is not even a unique topological specification. In the case of a free-boundary, low shear, low aspect ratio equilibrium, for example, the instability seems to change continuously from an $m=1$ to an $m=2$ mode as the q-value is raised, as illustrated in fig. 7.3. The

Toroidal Instabilities

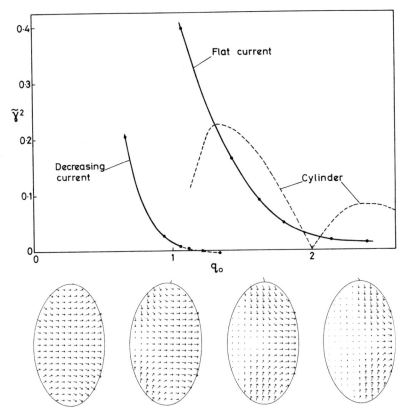

Fig. 7.3
Growth rates and eigenfunctions for free-boundary instabilities in a tokamak ($b/a = 1.6$, $R/a = 2.5$) with "flat current" ($p'(\psi)$ uniform, $II' = 0$, $q_{\text{edge}}/q_{\text{axis}} = 1.7$) and "decreasing current" ($p'(\psi) \sim \psi$, $II' = 0$). From J. A. Wesson and A. Sykes, IAEA Tokyo Conference, *1*, 449 (1974).

new structure gradually moves out from the inner edge of the torus. We shall see that the distinction between the straight cylinder and toroidal instabilities becomes greater as the poloidal beta is raised and the instability crowds into the outer edge of the torus.

The effect of toroidicity on the growth rate can be seen by taking a given instability in a straight cylinder and gradually bending the cylinder into a torus. For low beta plasmas, the effect of increasing toroidicity is to gradually reduce the growth rate. The growth rates are reduced more for long wavelengths, which sample a larger part of the curvature, than for shorter wavelengths.

7.5 SUMMARY

Most of the section on flux coordinates is intended as a summary of useful formulas for reference. It should be kept in mind that there are many possible flux coordinates. Hamada coordinates are the one particular choice in which magnetic field lines and current density lines appear straight and the Jacobian is normalized to unity everywhere.

The Mercier stability criterion is a necessary condition for stability which is derived by considering perturbations localized arbitrarily close to any rational surface within the plasma. The formal statements of the Mercier criterion (7.2.49) or (7.2.46) are exact for any equilibrium with nested flux surfaces and shear. A special evaluation of the Mercier criterion for a low beta, large aspect ratio, circular toroidal plasma (7.3.1) yields the necessary stability criterion $q \geq 1$ at the magnetic axis. Another approximation for an axisymmetric toroid with arbitrary cross section and poloidal beta yields (7.3.2) close to the magnetic axis.

On large-scale instabilities, toroidicity has the effect of mixing poloidal harmonics, especially for the perturbed poloidal magnetic field. At low beta, toroidicity tends to reduce the growth rate of the longer wavelength instabilities.

7.6 REFERENCES

The derivation of the Mercier criterion used here as well as a clear description of Hamada coordinates can be found in

J. M. Greene and J. L. Johnson, *Physics of Fluids*, 5, 510–517 (1962).

Additional physical insight is provided by

J. M. Greene and J. L. Johnson, *Plasma Physics*, 10, 729–745 (1968).

An alternative derivation is given in considerable detail by

C. Mercier and H. Luc, *The MHD Approach to the Problem of Plasma Con-*

finement in *Closed Magnetic Configurations*, EUR 5127e, Commission of the European Communities, Luxembourg, 1974.

References can be found in these papers to earlier derivations by Mercier, Bineau, and others. The form of the Mercier criterion (7.2.49) most widely used in computational work is derived by

L. S. Solov'ev, *Sov. Phys. JETP*, 26, 400–407 (1968).

An excellent short and long review of analytic toroidal stability conditions is given by

L. S. Solov'ev, *Soviet Atomic Energy*, 30, 14–21 (1971).

L. S. Solov'ev, *Reviews of Plasma Physics*, 6, 239–331 (New York: Consultants Bureau, 1976).

More information on flux coordinates can be found in

L. S. Solov'ev and V. D. Shafranov, *Reviews of Plasma Physics*, 5, 11–27, (New York: Consultants Bureau, 1970).

S. Hamada, *Nuclear Fusion*, 2, 23–37 (1962).

G. Bateman, *Nuclear Fusion*, 13, 227–238 (1973).

G. O. Spies and D. B. Nelson, *Physics of Fluids*, 17, 1879–1884 (1974).

Among the dozens of special evaluations of the Mercier criterion, the ones cited in section 7.3 are

V. D. Shafranov and E. I. Yurchenko, *Sov. Phys. JETP*, 26, 682–696 (1968).

D. Lortz and J. Nührenberg, *Nuclear Fusion*, 13, 821–827 (1973).

The corresponding sufficient criterion is derived by

D. Lortz, *Nuclear Fusion*, 13, 817–819 (1973).

Among the many analytic treatments of large-scale toroidal instabilities, the ones cited in section 7.4 are

A. A. Ware and F. A. Haas, *Physics of Fluids*, 9, 956–964 (1966).

A. A. Ware, *Physics of Fluids*, 7, 2006–2011 (1964); *Phys. Rev. Letters*, 26, 1304–1307 (1971).

E. A. Frieman, J. M. Greene, J. L. Johnson, and K. E. Weimer, *Physics of Fluids*, 16, 1108–1125 (1973).

M. N. Bussac, R. Pellat, D. Edery, and J. L. Soule, *Phys. Rev. Letters*, 35, 1638–1641 (1975).

A number of good papers giving computational results on large-scale instabilities in toroidal geometry happen to be collected together at the IAEA Tokyo Conference, *1*, 429–514 (1974). In particular the results given in section 7.4 come from

W. Schneider and G. Bateman, IAEA Tokyo Conference, *1*, 429–438 (1974).

J. A. Wesson and A. Sykes, IAEA Tokyo Conference, *1*, 449–461 (1974).

D. Berger, L. C. Bernard, R. Gruber, and F. Troyon, IAEA Berchtesgaden Conference, CN-35/B11-4 (1976).

In addition to these papers, the computer codes are described in

G. Bateman, W. Schneider, and W. Grossmann, *Nucl. Fusion*, *14*, 669–683 (1974).

A. Sykes and J. A. Wesson, *Nuclear Fusion*, *14*, 645–648 (1974).

The Princeton eigenvalue code is described by

R. C. Grimm, J. M. Greene, and J. L. Johnson, *Methods in Computational Physics*, *16*, 253–280 (New York: Academic Press, 1976).

For references on the Lausanne code, see Berger et al., cited above.

8
High Beta Tokamaks

Beta is a measure of the plasma energy density compared to the magnetic field energy density. The usual working definition for the average beta is

$$\langle \beta \rangle \equiv \langle p \rangle \Big/ \frac{1}{2\mu} B_{\varphi 0}^2 \qquad (8.0.1)$$

where $\langle p \rangle$ is the volume averaged plasma pressure and $B_{\varphi 0}$ is the vacuum toroidal magnetic field at the geometric center of the plasma.

Why do we want high beta for controlled thermonuclear fusion? The general argument is that it is expensive to produce a strong magnetic field and $\langle \beta \rangle$ is a measure of how efficiently the magnetic field is being used to confine the plasma. A more specific argument uses the observation that the thermonuclear power produced by a Deuterium-Tritium plasma scales like

$$n_D n_T \overline{\sigma v} \simeq 10^{-18} \frac{\text{cm}^{-3}}{\text{sec(keV)}^2} T_{DT}^2 [\text{keV}] n_D n_T [\text{cm}^{-3}]$$

assuming $10 \text{ keV} < T_e = T_D = T_T < 20 \text{ keV}$ and $n_D = n_T = \frac{1}{2} n_e$. It follows that the power output averaged over the plasma volume is proportional to

$$(\beta^*)^2 \equiv \langle p^2 \rangle \Big/ \left(\frac{1}{2\mu} B_{\varphi 0}^2\right)^2 \tag{8.0.2}$$

for a given magnetic field. Several conclusions can be drawn from this observation: (1) it is the plasma pressure, rather than the temperature or the density alone, that determines the power output; (2) a peaked pressure profile gives more power output than a broad profile, for the same average pressure; and (3) high $\langle \beta \rangle$ gives more power output than low $\langle \beta \rangle$ for the same profile shapes and magnetic field strength.

There have been many ways proposed to raise $\langle \beta \rangle$ or β^* in tokamaks. Some of the currently popular suggestions are

1. Elongate the cross section of the plasma so that the poloidal magnetic field can be raised;
2. Raise β_{pol} by heating the plasma rapidly so that poloidal currents are induced;
3. Surround these high-beta plasmas with a region with large force-free currents to improve the plasma stability; .
4. Broaden the toroidal current profile; and
5. Build tokamaks with low aspect ratio.

The first two methods will be discussed in the remainder of this chapter. Not much is known yet about the third method, although research is in progress. Broad current profiles produced by increasing the toroidal current in the plasma are accompanied by an enhanced level of unstable activity and an experimentally observed deterioration in the energy confinement time, as will be described in chapter 11. The choice of aspect ratio for tokamaks is usually an engineering compromise between the physics advantages of low aspect ratio and the need for accessibility as well as space to put the transformer and structural elements through the center of the torus. Most tokamaks have an aspect ratio between three and five.

8.1 ELONGATED CROSS SECTION

There is widespread controversy over the merits of building tokamaks or pinches with highly elongated cross section and questions of equilibrium and large scale stability lie at the center of this controversy. Many researchers agree that mild elongation and shaping would probably improve tokamak parameters, although until recently it was felt that such fine tuning was not yet worth the extra cost and effort. However, the idea of a highly elongated cross section ($b/a \gg 2$) has been heavily challenged on the basis of theoretical predictions that such plasmas would be grossly unstable at any q-value and that it would be impossible to hold the elongation once the current profile became peaked near the magnetic axis. Several of the experiments seem to defy these predictions of instability, although a variety of problems have prevented the experiments from giving conclusive evidence. I will try to give both sides of the story in this chapter.

The concept of building toroidal devices with elongated cross section was developed independently for a number of different reasons:

One rationale is based on the idea that gross stability and confinement are determined by the q-value alone, independent of the shape of the cross section. As the plasma is elongated with fixed q-value, the poloidal magnetic field must be increased in order to traverse the longer poloidal circumference with the same winding number. This stronger poloidal field confines a higher pressure and therefore a higher beta plasma. Also, with fixed minor radius, there is a correspondingly larger toroidal current density and therefore more Ohmic heating. Using the simple uniform current, elliptical cross-section model developed in section 4.5, the reader can demonstrate that both the poloidal magnetic field and toroidal current density increase like b/a (for large elongations, $b/a \gg 2$) and that the average beta and total toroidal current scale like $(b/a)^2$ as the plasma is elongated. It can be shown (Artsimovich and Shafranov (1972)) that there is stronger Ohmic heating, less diffusion, reduced heat loss, the possibility of a natural divertor near the tips of the elongation, and many other possible advantages.

On the basis of arguments similar to these, a series of shock heated experiments were built with highly elongated cross section, starting with Pharos at NRL (1968), TESI and TENQ at Jülich (1971), the Belt Pinches at Garching (1971), and the Lausanne Belt Pinch (1975). (See the IAEA or European conferences for these years.) Because of their inherently large inductance, it was hard to make the shock fast enough and hot enough to avoid rapid cooling due to impurity radiation. Also, the slow shocks led to excessive compression away from the walls so that it was often not possible for the current in the external coils to hold the highly elongated shape and the plasma rapidly contracted to a more circular cross section (as predicted in section 4.6). For these reasons, the experimental observation of overall stability when $q_{\text{edge}} \gtrsim 3$ (Gruber and Wilhelm (1976); Krause (1975)) has been the subject of debate.

Question 8.1.1

Apply the Suydam criterion to a thin cylindrical shell plasma (pressure peaked within the shell and zero outside). In addition to a toroidal current, suppose there is a poloidal current that runs parallel to the axis of the cylinder—an equal amount running in opposite directions through the inner and outer parts of the shell. Does the Suydam criterion impose a limit on the poloidal beta? Are the instabilities concentrated near the inner or the outer edge of the shell?

By 1968, Dr. Tihiro Ohkawa conceived of the elongated cross section as a multipole configuration with plasma currents replacing the imbedded multipole coils. Multipole devices have inherently high shear and a region of minimum average magnetic field strength. It is experimentally observed that they have lower fluctuation levels and longer confinement times than similar devices without internal multipoles. Ohkawa's concept lead to the Doublet series of experiments at General Atomic.

Question 8.1.2 (Ohkawa (1968))
What is the shape of a plasma with uniform current J_0 in the presence of a quadrupole (I_2) and an octopole (I_4) field?

$$\psi = -\pi J_0(x^2 + y^2) + I_2(x^2 - y^2) - I_4(x^4 - 6x^2y^2 + y^4)$$

How strong do the currents have to be to produce two o-type magnetic axes? What is the maximum width of the confined plasma?

Artsimovich and Shafranov (1972) listed the arguments for building a tokamak with an elongated D-shaped cross section. The first tokamak constructed in this form was the Finger Ring tokamak T-9. A small amount of triangularity, with the point facing away from the center line of the toroid, helps to satisfy the Mercier necessary and the Lortz sufficient stability criteria, as discussed in section 7.3.

It was shown in sections 4.5 and 4.6 that it is better to squeeze the plasma between walls of current running opposite to the plasma current rather than to pull on it with currents running parallel to the plasma current. There remains the problem that the walls must be close to the plasma and the current profile must be broad within the plasma in order to maintain this elongation. This is a problem for shock-heated devices because it is hard to avoid compressing the plasma as it is being shock heated. Very fast, high-voltage shocks must be used and the greater inductance of devices with elongated cross section does not lend itself to fast shocks. This equilibrium problem leads to strong axial contraction in shock heated devices as the elongated plasma

High Beta Tokamaks

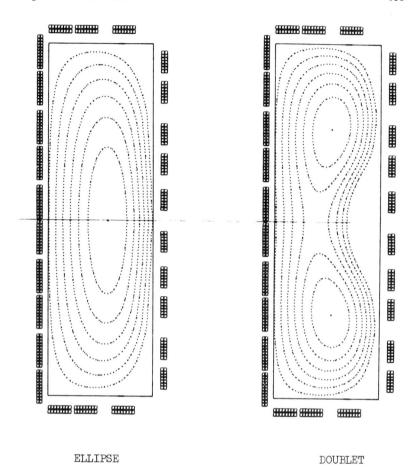

Fig. 8.1
Flux surfaces of an elongated plasma compared with a Doublet configuration. Also shown are cross sections of the coils used to shape the plasma. From the Doublet III design proposal by the General Atomic Company Fusion Staff, courtesy of EPRI.

contracts down to a more circular cross section. Rapid axial contraction of a plasma that had broken up into a collection of intertwined filaments was experimentally observed by Düchs, Dixon, and Elton (Garching High Beta Conf., 1972) in the hard-core Θ pinch at NRL. The process of axial contraction was studied in detail by Krause (1975) on the Belt Pinch.

Axial contraction is also a problem for Ohmically heated devices such as tokamaks because transport processes tend to make the current concentrate near the magnetic axis and away from the walls. As the current concentrates in the Doublet configuration, the plasma tends to shrink into "droplets," which are two separated plasma rings. By means of a delicate manipulation of the external currents in an active feedback arrangement, the droplets can be pushed back together to try to maintain the Doublet configuration. Then there is the danger that the rings will coalesce into a single ring with nearly circular cross section carrying too much current for the Kruskal-Shafranov stability condition. Since the profiles change on a slow transport time scale, there is a realistic hope that the equilibrium problem can be controlled by feedback.

Even if a way were found to control the profile and the shape of these elongated equilibria, there are more problems predicted for the large-scale stability of these configurations. First, let us consider the general effect of high beta on MHD instabilities.

8.2 HIGH BETA INSTABILITIES: SURFACE CURRENT MODEL

The most extensive theoretical study of the stability of high beta tokamaks has been carried out by Freidberg, Haas, Marder, Grossmann, and Goedbloed (1973–1976) at Los Alamos. For much of this work they used a surface current model to consider a wide variety of shapes and conditions. Their work indicates that the q-value is not the relevant parameter for high beta stability. Rather, there is a *critical beta* above which there is no stability at any q-value. For betas larger than this critical value, the instabilities observed are generally called *ballooning modes* because they force the plasma to bulge out most where the destabilizing curvature is the strongest. In order to study this kind of instability in its simplest form, consider the expression for δW in the surface current model.

It should be recalled from section 4.6 that in a purely surface current equilibrium the pressure is uniform within the plasma, $\nabla p^0 = 0$, and

High Beta Tokamaks

there is no current, $\mathbf{J}^0 = 0$, except at the surface of the plasma. It follows that the contribution to δW from within the plasma (5.4.2) reduces to

$$\delta W_F = \tfrac{1}{2} \int_{\text{plasma}} d^3x \left\{ \frac{1}{\mu} |\mathbf{B}^1|^2 + \Gamma p^0 |\nabla \cdot \boldsymbol{\xi}|^2 \right\} \tag{8.2.1}$$

where the perturbed magnetic field $\mathbf{B}^1$ is related to the displacement vector by (5.2.5)

$$\mathbf{B}^1 = \nabla \times (\boldsymbol{\xi} \times \mathbf{B}^0). \tag{8.2.2}$$

The contribution from the perturbation of the surface of the plasma (5.4.6) reduces to

$$\delta W_S = \tfrac{1}{2} \int d\mathbf{S} \cdot \left[\nabla \frac{1}{2\mu} (B^0)^2 \right] (\hat{\mathbf{n}} \cdot \boldsymbol{\xi})^2 \tag{8.2.3}$$

where $[\![\,]\!]$ stands for the jump from the inside to the outside of the surface. Finally, the contribution from the exterior vacuum region is (5.4.5)

$$\delta W_{\text{vac}} = \tfrac{1}{2} \int_{\text{vacuum}} d^3x \frac{1}{\mu} |\mathbf{B}^1|^2. \tag{8.2.4}$$

For determining the marginal stability condition, the perturbation that minimizes δW is needed. The contribution (8.2.1) within the plasma is minimized by an incompressible flow, $\nabla \cdot \boldsymbol{\xi} = 0$, and a vacuum perturbed magnetic field, $\nabla \times \mathbf{B}^1 = 0$. The contributions δW_F and δW_{vac} then reduce to the same form, which is most easily computed by setting

$$\mathbf{B}^1 \equiv \nabla \phi, \qquad \nabla^2 \phi = 0 \tag{8.2.5}$$

and then using Gauss's theorem to write

$$\int d^3x |\mathbf{B}^1|^2 = \int d\mathbf{S} \cdot \nabla \phi \phi. \tag{8.2.6}$$

In this way all the contributions to δW are reduced to surface integrals. Note that the scalar potential ϕ is generally not single valued in multiply connected regions.

Question 8.2.1

Is an incompressible displacement vector in (8.2.2) compatible with a vacuum magnetic field perturbation within the plasma?

In order to evaluate the surface contribution (8.2.3), break $(B^0)^2$ into the square of the poloidal and the toroidal field components separately. Since each is a vacuum field, $\nabla \times \mathbf{B} = 0$, we can write

$$\tfrac{1}{2}\nabla B^2 = \mathbf{B}\cdot\nabla\mathbf{B} = B^2\hat{\mathbf{B}}\cdot\nabla\hat{\mathbf{B}} + \hat{\mathbf{B}}\mathbf{B}\cdot\nabla|\mathbf{B}|.$$

The first term is the familiar curvature term, but now written for $\mathbf{B}^0_{pol}$ and $\mathbf{B}^0_{tor}$ separately, and the second term drops out of δW_S because $\mathbf{B}^0$ is tangent to the plasma surface. Using the fact that $\mathbf{B}^0_{pol} = 0$ within the plasma (for axisymmetric equilibria), (8.2.3) becomes

$$\delta W_S = -\frac{1}{2\mu}\int d\mathbf{S}\cdot\left\{\kappa_{pol}(B^0_{pol})^2_{ext} - \frac{\hat{\mathbf{R}}}{R}[\![(B^0_{tor})^2]\!]\right\}(\hat{\mathbf{n}}\cdot\boldsymbol{\xi})^2 \qquad (8.2.7)$$

where $\kappa_{pol} = \hat{\mathbf{B}}_{pol}\cdot\nabla\hat{\mathbf{B}}_{pol}$ and $\kappa_{tor} = -\hat{\mathbf{R}}/R$. These contributions to δW_S are locally destabilizing if either radius of curvature points into the plasma, and stabilizing wherever the radius of curvature points away from the plasma. Concave surfaces are stabilizing while convex surfaces produce a destabilizing contribution.

In order to illustrate the destabilizing effect of high beta, consider the surface current model for a straight circular cylindrical plasma. The perturbed magnetic field inside the plasma is given by

$$\mathbf{B}^1 = \nabla\phi, \quad \phi = -B^0_{z\,int}\xi_a\frac{I_m(kr)}{I'_m(ka)}e^{i(m\theta-kz)} \qquad (8.2.8)$$

so that the contribution to the potential energy is

$$\delta W_F = \frac{\pi}{2\mu}\xi_a^2\frac{(ka)^2}{m}(B^0_{z\,int})^2 \qquad (8.2.9)$$

where $B^0_{z\,int}$ is the B_z field within the plasma. The poloidal curvature is $\hat{\mathbf{r}}/a$ and the toroidal curvature is zero, so that (8.2.7) becomes

$$\delta W_S = -\frac{\pi}{2\mu}\xi_a^2(B^0_{pol})^2. \qquad (8.2.10)$$

The exterior vacuum contribution to δW is given by (6.4.2). Now define beta to be the ratio of the plasma pressure to the exterior magnetic energy

$$\beta \equiv \frac{2\mu p}{B^2_{z\,ext} + B^2_{pol}} = 1 - \frac{B^2_{z\,int}}{B^2_{z\,ext} + B^2_{pol}} \qquad (8.2.11)$$

and use the equilibrium relation

$$2\mu p = B^2_{pol} + B^2_{z\,ext} - B^2_{z\,int} \qquad (8.2.12)$$

together with the q-value per unit wavelength at the plasma surface

$$q = kaB_{z\,\text{ext}}/B_{\text{pol}} \tag{8.2.13}$$

to write

$$\delta W = \frac{\pi}{2\mu}\xi_a^2 B_{\text{pol}}^2 \left\{ \underbrace{-1 + \frac{(m-q)^2}{m}}_{\text{surface}} \underbrace{\frac{1+(a/r_{\text{wall}})^{2m}}{1-(a/r_{\text{wall}})^{2m}}}_{\text{vacuum}} \right.$$

$$\left. + \underbrace{\frac{1-\beta}{m}q^2[1+(ka/q)^2]}_{\text{interior}} \right\}. \tag{8.2.14}$$

For long wavelengths, $ka \ll 1$, with the wall far from the plasma, $a/r_{\text{wall}} \ll 1$, the plasma is unstable for

$$\beta > 1 - \frac{m-(m-q)^2}{q^2}. \tag{8.2.15}$$

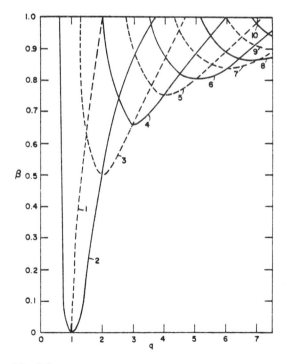

Fig. 8.2
Stability diagram β vs. q for the surface current model of a circular cylindrical plasma. The plasma is unstable above each curve to the poloidal mode number indicated. From B. M. Marder, *Phys. Fluids*, *17*, 447 (1974).

As shown in fig. 8.2, the envelope of the unstable region starts at $q = 1$ for low beta but rapidly moves to large q-values for high beta. For any q-value greater than one there is a *critical beta* for instability. Note that the beta can be changed independently of q because of the jump in B_z.

Question 8.2.2
Is this high beta instability easily stabilized by a wall or by finite wavelength $(ka \sim \mathcal{O}(1))$ effects?

If the plasma is elongated in the form of an ellipse this pressure driven instability gets much worse. While the poloidal magnetic field remains uniform around the surface in this straight cylinder geometry, the curvature increases like b/a^2 at the tips of the ellipse and decreases like a/b^2 along the sides. As the plasma is elongated, this increased curvature at the tips drives the instability at progressively lower beta, as shown in fig. 8.3. The form of the perturbation along the surface of

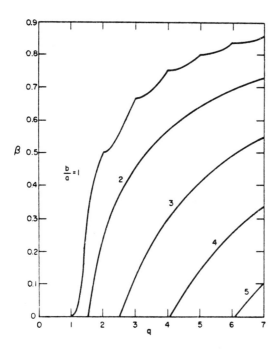

Fig. 8.3
The configuration in fig. 8.2 becomes more unstable as the cross section becomes an elongated ellipse. From B. M. Marder, *Phys. Fluids*, *17*, 447 (1974).

the plasma and the stabilizing magnetic energy terms were computed by Marder (1974); they are hard to estimate analytically. However, it is clear from the nature of the driving terms that the instability tends to blow out the ends of the ellipse, leaving the edges essentially untouched. It is also clear that the situation can be improved somewhat by reducing the curvature at the ends. For example, Marder considered a racetrack geometry—two straight sections connected by semicircles—and found some increase in the critical beta. For any given q-value, however, the poloidal magnetic field increases with elongation so that the destabilizing term becomes stronger than the interior stabilizing term even with optimum curvature. Elongation, therefore, gives only modest improvements on the critical beta in a straight cylinder.

The corresponding toroidal cases were worked out in the large aspect ratio limit by Freidberg and Haas (1974), and for any aspect ratio by Freidberg and Grossmann (1975). When $\beta \sim \mathcal{O}(\varepsilon)$, $\varepsilon \equiv R/a$, so that the jump in the toroidal field confines the plasma pressure more than the poloidal field, the last term in (8.2.7) provides a large destabilizing contribution along the outer edge of the toroid, in spite of the fact that the toroidal curvature is generally smaller than the poloidal curvature. For this reason the effect of toroidicity is generally destabilizing at high beta. Toroidicity also affects the stabilizing magnetic energy terms in a way that does not easily lend itself to interpretation.

Some large aspect ratio toroidal results are summarized in fig. 8.4. The maximum stable beta for a circular surface-current torus is found to be $\beta = 0.21 \ a/R$, which is much lower than the corresponding straight circular cylindrical case. With vertical elliptical elongation, the maximum critical beta rises to $\beta = 0.37 \ a/R$ as the elongation is increased to $b/a = 2.2$ and then falls off rapidly as the elongation is increased further.

When finite aspect ratio toroids are considered, the results are even more pessimistic. The results for a circular torus are presented in fig. 8.5. Note that the large aspect ratio limit $R/a \to \infty$ is at the lower left corner of this plot. Freidberg and Grossmann (1975) also give corresponding results for an elliptical and a Doublet elongation. The highest critical beta predicted for this surface-current Doublet model at fixed aspect ratio $R/a = 2.44$ is $\beta \simeq 0.13$ for $b/a \simeq 4$.

Of course this surface current model is not intended to make quantitative predictions about experiments—it only points out the trends. Much less extensive theoretical work has been done with distributed current models for plasmas with elongated cross section. The analytic

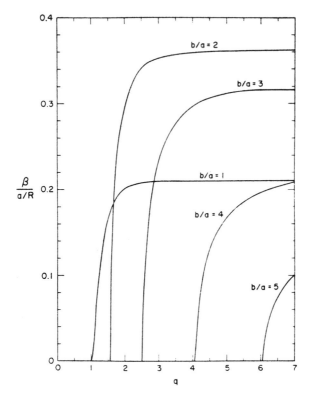

Fig. 8.4
Stability diagrams for toroidal surface-current plasmas with vertically elongated elliptical cross sections in the large aspect ratio limit. Plasma is unstable above each curve. From J. P. Freidberg and F. A. Haas, *Phys. Fluids*, **17**, 440 (1974).

treatment by Laval, Pellat, and Soulé (1974) on kink modes in elliptical plasmas with $\beta_{pol} \sim \mathcal{O}(1)$ indicates that a peaked current profile tends to suppress modes with higher poloidal harmonics, so that the marginal stability condition returns closer to $q = 1$. For example, if the equilibrium profile is determined by

$$J_{tor}(\psi) \sim \psi$$

giving a roughly parabolic current distribution, the high-m modes are unstable only in narrow intervals of q-value at the edge when $b/a \gg 1$

$$|nq_{edge} - m| < m \exp\left\{-\frac{4m}{\pi^2}\left(\frac{b}{a}\right)^2\right\}$$

where, here, m refers to the poloidal Fourier harmonic in the angle

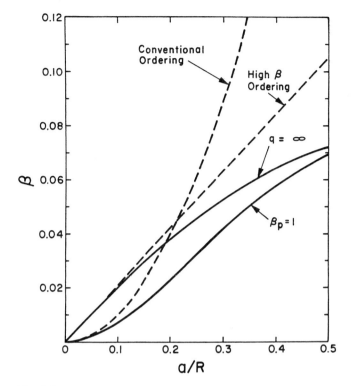

Fig. 8.5
Stability diagrams for a toroidal surface-current plasma with circular cross section. Solid curves are computed and the dashed curves are from analytic approximations. From J. P. Freidberg, J. P. Goedbloed, W. Grossmann, and F. A. Haas, IAEA Tokyo Conference, *1*, 505 (1974).

variable of elliptic coordinates. On the other hand, if the current density is uniform, the bands of unstable q-value broaden out and overlap completely as b/a is increased.

A similar overlap of instabilities is observed by Bateman, Schneider, and Grossmann (1974) in a computer study of fixed-boundary instabilities in a straight cylinder with rectangular cross section and $J_{\text{tor}} = J_0 \psi$. Each instability is observed to consist of a number of velocity vortex cells rolling off each other and helically twisted down the cylinder. As the cross section is elongated, the mode number can still be uniquely identified as half the number of vortex cells or as the number of positive peaks of perturbed pressure. Although the longitudinal current density and beta increase with elongation for the marginal point of each mode taken separately, the domains of instability for a succession of higher modes become broader and overlap

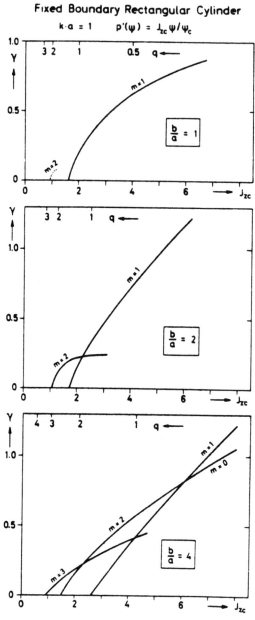

Fig. 8.6
Growth rate as a function of central current density for fixed-boundary instabilities in a straight cylinder with rectangular cross section. From G. Bateman, W. Schneider, and W. Grossmann, *Nuclear Fusion*, *14*, 669 (1974).

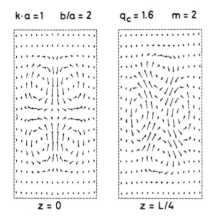

Fig. 8.7
Flow pattern for an $m=2$ fixed-boundary mode in a rectangular straight cylinder at cross sections one quarter of a wavelength apart. From G. Bateman, W. Schneider, and W. Grossmann, *Nuclear Fusion*, *14*, 669 (1974).

as the plasma is elongated, as shown in fig. 8.6. What is worse, the instabilities with higher mode number appear to become more virulent at higher elongations. For example, the $m=2$ mode shown in fig. 8.7 is no longer a benign localized instability. Still, it remains to be seen how the nonlinear evolution of these instabilities rearranges the plasma under realistic experimental conditions.

8.3 FLUX CONSERVING TOKAMAKS

An alternative to elongating the plasma cross section to achieve high beta is to heat the plasma rapidly and then rely on poloidal diamagnetic currents to confine the additional pressure. These poloidal currents are induced as the heated plasma expands slightly. For many years, it was believed that the plasma pressure in tokamaks would be limited by a simple equilibrium consideration: As the pressure is increased, the vertical magnetic field needed to prevent the plasma from expanding along the major radius (see section 4.6) must be raised and will cancel the poloidal field at the inner edge of the toroid when $\beta_{\text{pol}} \lesssim 1/\varepsilon$, $\varepsilon \equiv a/R$, and hence $\beta \sim \beta_{\text{pol}} B_{\text{pol}}^2/B_{\text{tor}}^2 \lesssim \varepsilon/q^2$. The resulting separatrix would reduce plasma confinement and cause the plasma to shrink. However, if the plasma were heated on a time scale faster than the resistive diffusion time scale (but slower than the Alfvén transit time scale) the magnetic fluxes would be frozen into the plasma and therefore, a separatrix could not move in to shrink the cross section. This concept, called the *flux conserving tokamak* (FCT) was developed analytically by Clarke and Sigmar (1977) and computationally by Dory and Peng (1977). There now appears to be no equilibrium limit on the volume-averaged beta that can be achieved in a tokamak.

To theoreticians, the flux conserving tokamak sequence is one of the possible paths leading to equilibria with arbitrarily high beta. For many years, researchers computed tokamak equilibria by making convenient choices for the functions $p(\psi)$ and $I(\psi)$ ($\equiv RB_\varphi$) which appear in the Grad-Shafranov equation (4.4.10). When pathologies developed at high beta, such as a separatrix or a reversed toroidal current at the inner edge of the toroid (Callen and Dory (1972)), it was never clear how to choose $p(\psi)$ or $I(\psi)$ to improve the equilibria. In a flux conserving sequence of equilibria, however, a given choice is made for the function $q(\psi)$ ($\equiv d\psi_{tor}/d\psi_{pol}$) and then any series of choices can be made for $p(\psi)$. The source function $I(\psi)$ needed for the Grad-Shafranov equation is then determined iteratively from the relation

$$q(\psi) = \frac{I(\psi)}{2\pi} \oint \frac{dl}{|B_{pol}|} \frac{1}{R^2} \tag{8.3.1}$$

where $\oint dl$ is the line integral around a flux surface at any cross section. To derive this relation, note that the differential element of area between flux surfaces ψ and $\psi + d\psi$ is $\oint dl \frac{d\psi}{|\nabla\psi|}$. It then follows that

$$q(\psi) = \frac{d\psi_{tor}}{d\psi_{pol}} = \frac{d}{d\psi_{pol}} \int dS \cdot \hat{\varphi} B_\varphi = \frac{1}{d\psi_{pol}} \oint dl \frac{d\psi_{pol}}{|\nabla\psi_{pol}|} \frac{I(\psi)}{R}$$

and hence (8.3.1). Iteration is required because $\oint (dl/|B_{pol}|)R^{-2}$ can be computed only after $\psi(R, y)$ is known.

To show that the functional form of $q(\psi)$ remains fixed during a flux conserving sequence, use the mathematical identity

$$\frac{d}{dt} \int dS \cdot B \equiv \int dS \cdot \left\{ \frac{\partial}{\partial t} B + v\nabla \cdot B - \nabla \times (v \times B) \right\} \tag{8.3.2}$$

and Faraday's law $\partial B/\partial t = \nabla \times (v \times B)$ to show that all magnetic fluxes are convected by a perfectly conducting fluid

$$\frac{d\psi}{dt} = 0 \quad \text{for any} \quad \psi \equiv \int dS \cdot B. \tag{8.3.3}$$

Then, as the pressure is pumped up (by heating or increasing density, for example) and the plasma rearranges itself to evolve from one equilibrium to another, it follows that the q-profile must convect along

High Beta Tokamaks

with the fluxes

$$\frac{d}{dt} q(\psi, t) = 0 \Rightarrow q = q(\psi). \tag{8.3.4}$$

Two examples of flux conserving tokamak equilibria are shown in fig. 8.8. When beta becomes comparable the reciprocal of the aspect ratio, $\varepsilon \equiv a/R$, the plasma pressure becomes confined mostly by poloidal diamagnetic currents crossed with the toroidal magnetic field, while the poloidal magnetic field serves mainly to center the plasma relative to the walls. It can be seen in fig. 8.8 that the central flux surfaces become shifted out toward the outer edge of the toroid. Also the toroidal current becomes concentrated into a crescent shape around the outer edge. A dead space, with low pressure and current density, develops near the inner edge of the plasma. For the most effective use of the plasma volume, this dead space should be minimized by forming

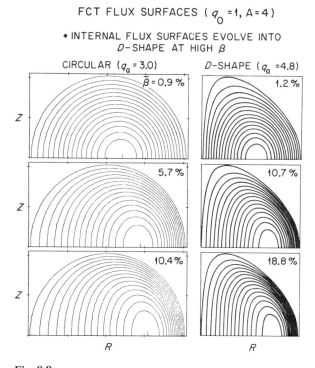

Fig. 8.8
Flux surfaces for two sequences of flux conserving tokamak equilibria. The toroidal center line is to the left. From R. A. Dory and Y-K. M. Peng, *Nuclear Fusion*, 17, 21 (1977).

the plasma into a D-shaped cross section. It can be seen in fig. 8.8 that the average beta is much higher in the D-shaped plasma even with a larger q-value at the plasma edge ($q_{\text{axis}} = 1$ in all cases shown). At low aspect ratio, the D shape seems to be the natural shape for a toroidal plasma and, as we shall see, it also seems to have the best stability properties.

Question 8.3.1
Suppose the plasma is surrounded by a set of high inductance toroidal field coils. As the plasma pressure is increased and poloidal diamagnetic currents form, how is the toroidal flux within the coils conserved?

Question 8.3.2
Consider a plane slab of current-carrying plasma bounded by perfectly conducting walls, with a perfectly conducting membrane in the middle. Suppose the membrane is shifted over closer to one of the walls. If the motion is flux conserving, how does the current density and the total current in the slab respond? Does the outward shift of the magnetic axis in a toroidal plasma have the same effect?

Question 8.3.3
Using a cause and effect logical sequence, can you demonstrate how plasma heating and subsequent expansion induces diamagnetic poloidal currents? What effect does the expansion have on the longitudinal current density?

Question 8.3.4
Consider a straight circular cylindrical plasma surrounded by a vacuum and a perfectly conducting wall. Suppose the plasma is heated and expands, conserving the q-value ($q = aB_z/RB_\theta$) at the edge of the plasma and both magnetic fluxes in the vacuum. Do these three constraints match at the plasma boundary?

It now appears that instabilities may set the limit on the maximum beta that can be confined in a tokamak. Ironically, the predicted instability beta limit is roughly the same as what was previously believed to be the equilibrium beta limit

$$\beta < \text{(Geometric factor)} \, \varepsilon/q^2, \quad \varepsilon \equiv a/R \qquad (8.3.5)$$

A simple rough derivation of this important stability criterion is given by Todd et al. (1977) and will be outlined here.

Consider the form of the energy principle given by (5.4.9)

$$\delta W = \tfrac{1}{2} \int d^3 x \left\{ \frac{1}{\mu} |\mathbf{B}_\perp^1|^2 + \frac{1}{\mu} |\mathbf{B}_\parallel^1 - \mu \mathbf{B}\boldsymbol{\xi} \cdot \nabla p/|B||^2 + \Gamma p |\nabla \cdot \boldsymbol{\xi}|^2 \right. $$
$$\left. + \mathbf{J} \cdot \hat{\mathbf{B}} \hat{\mathbf{B}} \times \boldsymbol{\xi} \cdot \mathbf{B}^1 - 2\boldsymbol{\xi} \cdot \nabla p \boldsymbol{\xi} \cdot \boldsymbol{\kappa} \right\} \quad (5.4.9)$$

where

$$\mathbf{B}^1 = \nabla \times (\boldsymbol{\xi} \times \mathbf{B}) = \mathbf{B} \cdot \nabla \boldsymbol{\xi}_\perp - \mathbf{B} \nabla \cdot \boldsymbol{\xi}_\perp - \boldsymbol{\xi}_\perp \cdot \nabla \mathbf{B}. \quad (8.3.6)$$

The minimum value of δW must be positive for stability. A simple minimizing procedure can be carried out in the following way: The third positive definite term in (5.4.9) can be eliminated by minimizing with respect to $\xi_\parallel$—the component of the displacement parallel to the equilibrium magnetic field—which does not appear in any of the other terms. Now split the perturbed magnetic field $\mathbf{B}^1$ into parallel and perpendicular components

$$\mathbf{B}_\parallel^1 = -\boldsymbol{\xi}_\perp \cdot \nabla |B| - |B| \nabla \cdot \boldsymbol{\xi}_\perp + \mathbf{B} \cdot \nabla \boldsymbol{\xi}_\perp \cdot \hat{\mathbf{B}} \quad (8.3.7)$$

$$\mathbf{B}_\perp^1 = \mathbf{B} \cdot \nabla |\xi_\perp| \hat{\xi}_\perp - |B| \boldsymbol{\xi}_\perp \cdot \nabla \hat{\mathbf{B}} + |\xi_\perp|(\mathbf{B} \cdot \nabla \hat{\xi}_\perp)_\perp \quad (8.3.8)$$
$$\text{neglect} \qquad \text{neglect}$$

where $\hat{\mathbf{B}}$ and $\hat{\boldsymbol{\xi}}_\perp$ stand for unit vectors and the terms to be neglected here are noted. Using the identity

$$\mu \boldsymbol{\xi} \cdot \nabla p/|B| = \boldsymbol{\xi}_\perp \cdot [-\nabla |B| + |B|\boldsymbol{\kappa}] \quad (8.3.9)$$

where $\boldsymbol{\kappa} \equiv \hat{\mathbf{B}} \cdot \nabla \hat{\mathbf{B}}$ is the curvature of the equilibrium field line, it can be shown that the second term in (5.4.9) can be eliminated by taking

$$\nabla \cdot \boldsymbol{\xi}_\perp \simeq -2\boldsymbol{\xi} \cdot \boldsymbol{\kappa}. \quad (8.3.10)$$

Essentially, this condition arises from minimizing δW with respect to the component of $\boldsymbol{\xi}$ in the magnetic surface but perpendicular to $\mathbf{B}$.

Finally, we consider equilibrium conditions under which the instability is primarily pressure driven (last term in (5.4.9)) so that the next to last term, which drives current-driven instabilities, can be neglected. The marginal stability criterion then comes from balancing the last term

$$-2\boldsymbol{\xi} \cdot \nabla p \boldsymbol{\xi} \cdot \boldsymbol{\kappa} \sim 2|p/a| \frac{1}{R} |\xi_\perp|^2 \quad (8.3.10)$$

against the stabilizing effect of the first term

$$\frac{1}{\mu}|B_\perp^1|^2 \simeq \frac{1}{\mu}(\mathbf{B}\cdot\nabla|\xi_\perp|)^2 \sim \frac{1}{\mu}|B/Rq|^2|\xi_\perp|^2. \tag{8.3.11}$$

When estimating the magnitude of $\mathbf{B}\cdot\nabla|\xi_\perp|$, it is assumed that $|\xi_\perp|$ is large only around the outer edge of the torus, along a length $1/Rq$ of the field line. The minor radius of the torus, a, is taken to be the scale length of the pressure gradient and the curvature of field line around the outer edge of the torus is roughly $1/R$ when the q-value is larger than one. The stability criterion (8.3.5) then follows by balancing these estimates of the terms in (5.4.9). It is clear that the instability will be strongest where the pressure gradient is largest, giving these modes their characteristic ballooning behavior. Note that the q-value on that surface passing through the point of maximum pressure gradient should be used in (8.3.5).

The marginally stable β scaling ε/q^2, as in (8.3.5), has been rigorously derived in the limit of large toroidal mode number ($n \gg 1$) by Dobrott, Nelson et al. (1977). In fig. 8.9, it can be seen how this scaling fits together with the stability criterion $q_{\text{axis}} > 1$ for the internal $m=1$ kink

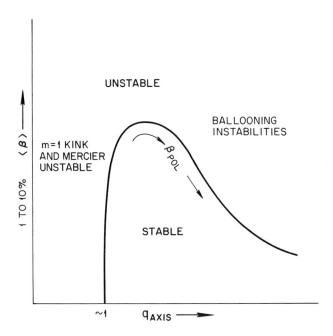

Fig. 8.9
Schematic stability diagram for a toroidal plasma. After D. B. Nelson.

High Beta Tokamaks

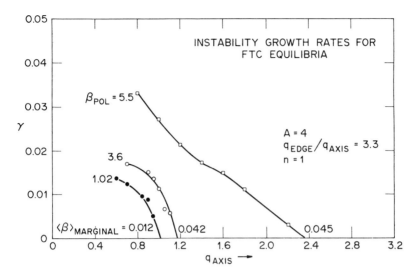

Fig. 8.10
Growth rate curves for the toroidal plasmas shown on the right in fig. 8.8 with increasing poloidal beta. From G. Bateman and Y-K. M. Peng, *Phys. Rev. Letters*, *38*, 829 (1977).

mode or the Mercier criterion (in a circular cross section). As poloidal beta is increased, moving along the marginal stability curve from the lower left, there is a sharp transition from the low beta stability criterion $q_{\text{axis}} > 1$ to the high beta criterion $\beta \lesssim \varepsilon/q^2$. The height of this curve and, therefore, the maximum stable beta that can be achieved, depend upon the shape and profile of the plasma. It is clear that low aspect ratio (low R/a) is good for ballooning mode stability. Some elongation and a D-shape further improve the maximum beta. A broad pressure and current profile is good for internal modes but is bad for free boundary modes, so that a compromise is needed for optimum stability. A force free region around the plasma appears to be good for stability, but once again a compromise is needed to avoid making the high beta core of the plasma too small. Shear has a mixed effect on ballooning modes but, as we shall see, too much shear enhances resistive instabilities.

As the poloidal beta is raised, there is a striking change in the growth rate curve (fig. 8.10) and the spatial structure (fig. 8.11) of the instabilities. At low beta the marginal stability eigenfunctions are localized near the magnetic axis, as illustrated in figs. 8.11 b and c for a typical $m = 1$, $n = 1$ instability. However, as β_{pol} is raised, the marginal stability eigenfunction moves out closer to the edge of the plasma and

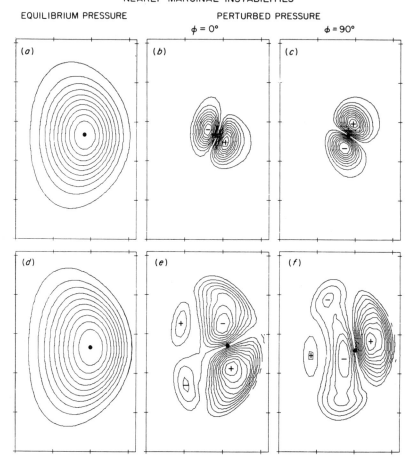

Fig. 8.11
Equilibrium pressure contours (a, d) and perturbed pressure contours at cross sections one quarter of a wavelength apart for instabilities near marginal points. Frames a, b, and c refer to the point in fig. 8.10 with $\beta_{pol} = 1.02$, $q_{axis} = 0.9$, and $\langle \beta \rangle = 1.56\%$. Frames d, e, and f refer to $\beta_{pol} = 3.6$, $q_{axis} = 1.1$, and $\langle \beta \rangle = 4.85\%$. From G. Bateman and Y-K. M. Peng, *Phys. Rev. Letters*, *38*, 829 (1977).

takes on a strong ballooning character, as shown in figs. 8.11 e and f. The perturbed pressure shown in these contour plots results from convection around vortex cells in the poloidal plane, illustrated in fig. 8.12 for a torus with circular cross section. The high beta eigenfunction consists of many poloidal harmonics which always constructively reinforce each other at the outer edge of the toroid and nearly vanish at the inner edge.

These differences between the low beta and high beta instabilities are expected to have a profound effect on the nonlinear development and consequences of the instabilities. The low beta instabilities are relatively benign under normal tokamak operating conditions. As we shall see in chapter 12, the resistive form of the internal $m=1$ low beta instability is believed to be responsible for the sawtooth oscillations observed in tokamaks; they churn around in the middle of the plasma, leaving the edge essentially untouched. However, because the high beta ballooning modes concentrate near the edge of the plasma, or wherever the pressure gradient is largest, they are potentially much more dangerous. Since ballooning modes are analogous to the Rayleigh-Taylor instability, they might be expected to establish Bénard convection cells (see chapter 10) which might greatly enhance plasma losses. Several tokamak experiments are now being designed to test this stability limit and determine the consequences.

8.4 SUMMARY

As the plasma pressure is raised, a new class of instabilities, called ballooning modes, are predicted to take over as the most dangerous and limiting MHD instabilities in tokamaks. The term ballooning mode refers to an instability that is not uniform along magnetic field lines. They cause the plasma to bulge out most where both the pressure gradient and curvature are strongest and point in roughly the same direction. Hence, they are typically strongest at the outer edge of a circular high beta toroidal plasma or at the tips of an elongated plasma. While the kink mode or Mercier stability criterion imposes a limit on the q-value in the plasma, such as $q \gtrsim 1$ at the magnetic axis of a large aspect ratio circular plasma, ballooning mode stability imposes a limit on the pressure or, its dimensionless form, the beta of the plasma, as in (8.3.5). This beta limitation depends sensitively on the geometry and profile of the plasma.

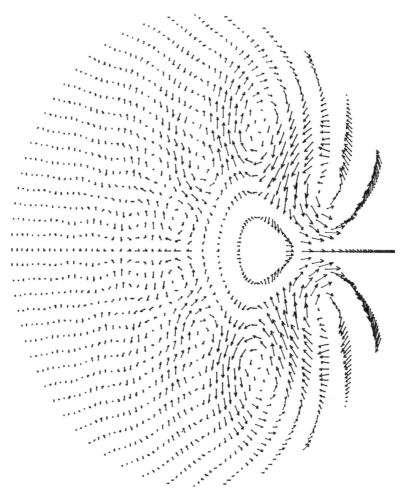

Fig. 8.12
Flow pattern for ballooning mode with $n = 3$ at $\beta^* = 3.0\%$. From A. M. M. Todd, M. S. Chance, J. M. Greene, R. C. Grimm, J. L. Johnson, and J. Manickam, *Phys. Rev. Letters*, *38*, 826 (1977).

8.5 REFERENCES

Much of the experimental work on elongated plasmas is published in the IAEA and the European conference proceedings. See also the topical conferences on pulsed high beta plasmas:

I: Los Alamos Report LA-3770 (1967),

II: Garching, IPP-Report 1/127 (1972), and

III: Culham; *Plasma Physics* supplement, Pergamon Press (1975).

Additional references from the text are

L. A. Artsimovich and V. D. Shafranov, *JETP Letters*, 15, 51–54 (1972).

O. Gruber and R. Wilhelm, *Nuclear Fusion*, 16, 243–251 (1976).

H. Krause, *Nuclear Fusion*, 15, 855–863 (1975).

T. Ohkawa and H. G. Voorhies, *Phys. Rev. Letters*, 22, 1275–1277 (1969).

T. Ohkawa, Report GA-8528 (1968).

Stability computation using the surface current model were reported by

J. P. Freidberg and F. A. Haas, *Phys. Fluids*, 17, 440–446 (1974).

B. M. Marder, *Phys. Fluids*, 17, 447–451, 634–639 (1974).

J. P. Freidberg and W. Grossmann, *Phys. Fluids*, 18, 1494–1506 (1975).

J. P. Freidberg, J. P. Goedbloed, W. Grossmann, and F. A. Haas, IAEA Tokyo Conference, 1, 505–514 (1975).

For the effect that elongation has on instabilities, not including ballooning modes, see

G. Laval, R. Pellat, and J. S. Soulé, *Phys. Fluids*, 17, 835–845 (1974).

G. Laval, *Phys. Rev. Letters*, 34, 1316–1320 (1975).

G. Bateman, W. Schneider, and W. Grossmann, *Nuclear Fusion*, 14, 669–683 (1974).

A. W. Allen, F. L. Cochran, G. C. Goldenbaum, and P. 'C. Liewer, *Phys. Rev. Letters*, 39, 404–407 (1977).

W. Grossmann, J. A. Tataronis, and H. Weitzner, *Phys. Fluids*, 20, 239–251 (1977).

R. L. Dewar, R. C. Grimm, J. L. Johnson, E. A. Frieman, J. M. Greene, and P. H. Rutherford, *Phys. Fluids*, 17, 930–938 (1974).

M. S. Chance, J. M. Greene, R. C. Grimm, and J. L. Johnson, *Nuclear Fusion*, 17, 65–84 (1977).

There are many additional papers in which the Mercier or a sufficient stability criterion is evaluated in noncircular cross-section geometry. For example,

L. S. Solov'ev, *Reviews of Plasma Physics*, 6, 239–331 (New York: Consultants Bureau, 1976).

D. Lortz and J. Nührenberg, *Nuclear Fusion*, 13, 821–827 (1973).

Y-K. M. Peng et al., *Phys. Fluids*, *21*, 467–475 (1978).

M. Okamoto, M. Wakatani, and T. Amano, *Nuclear Fusion*, *15*, 225–236 (1975).

F. Hernegger and E. K. Maschke, *Nuclear Fusion*, *14*, 119–121 (1974).

B. Coppi, R. Dagazian, and R. Gajewski, *Phys. Fluids*, *15*, 2405–2418 (1972).

For stability with respect to axisymmetric modes, see

F. A. Haas and J. C. B. Papaloizou, *Nuclear Fusion*, *17*, 721–728 (1977).

M. S. Chu, R. L. Miller, and T. Ohkawa, *Nuclear Fusion*, *17*, 465–472 (1977).

E. Rebhan and A. Salat, *Nuclear Fusion*, *17*, 251–261 (1977).

For more details on the flux conserving tokamak concept, see

J. F. Clarke and D. J. Sigmar, *Phys. Rev. Letters*, *38*, 70–74 (1977).

R. A. Dory and Y-K. M. Peng, *Nuclear Fusion*, *17*, 21–31 (1977).

J. D. Callen and R. A. Dory, *Phys. Fluids*, *15*, 1523–1528 (1972).

For recent work on ballooning mode theory in tokamaks with diffuse profile, see

A. M. M. Todd, M. S. Chance, J. M. Greene, R. C. Grimm, J. L. Johnson, and J. Manickam, *Phys. Rev. Letters*, *38*, 826–829 (1977).

G. Bateman and Y-K. M. Peng, *Phys. Rev. Letters*, *38*, 829–832 (1977).

A. Sykes, J. A. Wesson, and S. J. Cox, *Phys. Rev. Letters*, *39*, 757–760 (1977).

B. Coppi, *Phys. Rev. Letters*, *39*, 939–942 (1977).

D. Dobrott, D. B. Nelson, J. M. Greene, A. H. Glasser, M. S. Chance, E. A. Frieman, *Phys. Rev. Letters*, *39*, 943–946 (1977).

J. W. Connor, R. J. Hastie, and J. B. Taylor, *Phys. Rev. Letters*, *40*, 396–399 (1978).

9
Nonlinear Instability Theory

It used to be assumed that all macroscopic instabilities were harmful and that MHD theory should be used only to search for completely stable configurations. Since the marginal stability condition was then the only information of interest, theoretical research was concentrated almost exclusively on developing techniques for minimizing the potential energy. Until recently, few researchers bothered to look at the growth rate or structure of linear instabilities.

Now it is becoming clear that completely stable plasmas are not necessarily the best plasmas for controlled thermonuclear fusion. Fluctuations are routinely observed in tokamak plasmas in the form of Mirnov oscillations and sawtooth oscillations, as will be described in detail in chapter 11. It appears that the best operating conditions in tokamaks are observed when there is a balance between the power being put into the plasma and the enhanced losses due to unstable activity. To reduce the power level to the point where fluctuations can no longer be detected, as in fig. 1.8, is to operate with a relatively cold uninteresting plasma. Since it is evidently more important to learn to live with instabilities than to avoid them completely, there is now a great deal of interest in developing a nonlinear theory to predict the consequences of instabilities when they do occur.

In any event, just what is the nonlinear development of the instabilities we have been studying? Do they saturate at a finite amplitude?

Does their structure change dramatically from their appearance during the linear phase? Do they break up into fine scale structure? Do they carry the plasma to the wall? These are the questions we shall consider in this chapter.

9.1 NONLINEAR METHODS

There are at least five overlapping techniques used to study nonlinear MHD instabilities: computer simulation, bifurcation theory, singular perturbation analysis, the study of convection cells, and the study of fully developed turbulence. A brief history of each technique will be given in this section before a few of the most pertinent results are described in the next two sections.

Computer Simulation

The first attempted computer simulation of the nonlinear evolution of a large-scale instability was made by Roberts and Curtis with their "wriggling discharge model" in 1958. Then there was nothing for the next 15 years until Van Hoven and Cross (1973) studied the nonlinear evolution of a resistive tearing mode (see chapter 10) as a model for solar flares. All at once there were a series of nonlinear computer codes appearing in 1974 designed to follow the evolution of the ideal or resistive MHD equations long enough to study the effects of large scale instabilities. By the end of 1975 there were at least eight distinct research groups in the world with codes designed specifically to follow the nonlinear motion of instabilities and there were several more groups with codes designed to look for new three-dimensional equilibria using iteration procedures analagous to time evolution.

J. U. Brackbill (1976) started producing nonlinear results in the spring of 1974 with a highly sophisticated computer code using features long under development by fluid dynamicists at Los Alamos. Together with D. C. Barnes he has investigated shock heated plasmas such as the rotating theta pinch and models of the Scyllac device. Under these conditions the instability is observed to carry the plasma very close to the wall while the structure of the instability does not change appreciably from its linear form. The same code can be used to study the dynamic evolution of a stable plasma as it approaches equilibrium.

Starting in 1974, a series of nonlinear codes were developed under the direction of M. N. Rosenbluth at the Institute for Advanced Study in Princeton. A common feature of all these codes is the assumption of zero beta in the form of a large aspect ratio expansion. The equations are further simplified in most of the codes by considering only a straight circular cylindrical equilibrium and assuming that the perturbation is helically symmetric during its entire nonlinear evolution. This helical assumption is consistent for any given mode but it precludes the study of the interaction of several modes with different helicities. These assumptions lead to a set of equations for incompressible motion with the longitudinal magnetic field held fixed. In addition to reducing the number of variables and equations, these assumptions eliminate the fast magnetosonic and Alfvén time scales so that much longer timesteps can be taken while maintaining numerical stability and accuracy.

Some of the ideal MHD results by the Rosenbluth team are presented in the next section while their resistive instability results are presented in chapter 10. H. R. Strauss (1976) has extended this zero beta method to look at elongated cross sections, where the internal kink mode grows strongly. The helically symmetric assumption in a circular cylinder has also been used by Gerlach and Zueva (1974-1976) to study the effect of resistivity on MHD instabilities with finite beta and finite aspect ratio. Because of the greater accuracy that can be achieved by computing with these reduced sets of equations—especially when the problem is reduced to two dimensions by the helical assumption—this line of research has quickly proved to be very fruitful.

In an effort to study fixed-boundary instabilities, Hicks, Wooten, and Bateman at Oak Ridge developed a simple nonlinear code which uses an explicit leapfrog finite difference scheme on a Cartesian grid in a cylinder or toroid with rectangular cross section. Some of the results will be described in section 9.3. Wesson and Sykes (1976) at Culham added viscosity, resistivity, Ohmic heating, and diffusion to a similar code and observed a relaxation oscillation which looks like the sawtooth oscillations observed in soft X-ray diagnostics on tokamak experiments. Dnestrovski and coworkers in Moscow have written another variation of this code and have studied the coupling between $m=1$ and $m=2$ modes during their nonlinear evolution in a straight cylinder. A more sophisticated series of nonlinear codes, with all the effects of finite beta, resistivity, viscosity, and so forth, are being developed under the direction of J. Killeen at Livermore.

Bifurcation Theory

For parameters close to marginal stability, the basic idea is to look for a new deformed equilibrium in the neighborhood of the original equilibrium. A simple example of bifurcation was discussed in section 4.5. Here we are concerned with a straight cylindrical or axisymmetric toroidal equilibrium deforming into a new helical equilibrium. Usually the analysis is carried out by expanding the potential energy or the equation of motion to higher order in the displacement. This kind of analysis goes back to K. O. Friedrichs (1960) for short wavelength kink instabilities using a surface current model. The same analysis was extended to long wavelengths by Yeh (1973). Rutherford, Furth, and Rosenbluth (1971), in their search for the mechanism of the disruptive instability, found bifurcated equilibria for both kink and tearing modes in a straight cylinder with distributed current and a vacuum region. Kadomtsev (1963) and Kadomtsev and Pogutse (1973), (1974) suggest several elementary models to characterize the corkscrew and filamentary equilibria that result from free-boundary kink instabilities under certain conditions, as will be described in the next section.

Singular Perturbation Analysis

For the $m=1$ internal kink mode investigated by Rosenbluth, Dagazian, and Rutherford (1973), and for the nonlinear growth of the tearing mode investigated by Rutherford (1973), for example, the nonlinear terms are ignored everywhere except in a thin boundary layer around the mode rational surface. The nonlinear solution in this boundary layer is then matched onto the linear solution everywhere else.

Convection Cells

The classic example of an instability leading to steady state convection cells is the appearance of Bénard convection when a fluid is heated from below. This phenomenon is thoroughly described by Chandrasekhar (1961). There is evidence that convection cells make a large contribution to particle loss in certain plasma devices with low shear (see, for example, Harries (1970) and a number of papers presented at the IAEA Novosibirsk Conference (1968)). For recent theoretical work applicable to confined plasmas see Wobig (1972) and Maschke and Paris (1974).

Fully Developed Turbulence

There have been very few attempts to consider a macroscopic turbulent state in which a wide spectrum of MHD modes interact nonlinearly. A semi-empirical approach was outlined by Kadomtsev and Pogutse

Nonlinear Instability Theory

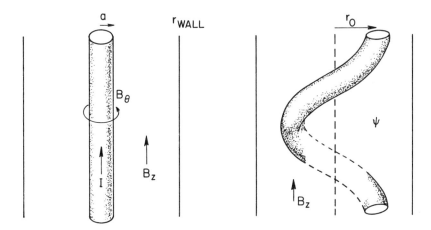

Fig. 9.1
Visualization of an $m=1$ kink mode to illustrate Kadomtsev's nonlinear stabilization mechanism.

(1967). Recently, a method developed by Kraichnan (1967) was applied to a two-dimensional MHD model by Montgomery and coworkers (for example, Fyfe and Montgomery (1975)). While the energy always cascades to shorter wavelengths in 3-D situations, they find that energy can cascade to both longer and shorter wavelengths in 2-D situations. The cascade of energy to the longer wavelengths produces vortex streets after a sufficiently long time. A low shear plasma in a strong magnetic field may behave more like the 2-D situation.

9.2 FREE-BOUNDARY INSTABILITIES

A very simple illustrative example of the nonlinear evolution of a kink instability was suggested by Kadomtsev (1963): Consider the surface current model for a thin straight cylindrical plasma separated from the wall by a vacuum. Let there be poloidal ($B_\theta = B_{\theta a} a/r$) and longitudinal ($B_z = B_{z0}$) magnetic field components in the vacuum region but no magnetic field inside the plasma. The poloidal flux between the plasma and the wall is

$$\psi_{\text{pol}} = \int dl \int dr\, B_\theta = aLB_{\theta a}\ln\frac{r_{\text{wall}}}{a}. \tag{9.2.1}$$

Now suppose the instability deforms the plasma into a helix with one turn over length L. The increment to the flux because the helix is wrapping around part of the longitudinal field, as illustrated in fig. 9.1, is

$$\delta\psi = \pi\xi^2 B_z \tag{9.2.2}$$

where ξ is the amplitude of the helical perturbation. If the poloidal flux is to be conserved during the perturbation, the longitudinal current and the poloidal field must decrease as the helix gets larger. A new equilibrium is reached when the longitudinal current within the plasma has decreased to zero, so that there is only the original uniform longitudinal magnetic field left. The radius ξ at which this happens is given by flux conservation

$$\delta\psi = \psi_{\text{original}} \Rightarrow \pi\xi^2 B_z = aLB_{\theta a}\ln\frac{r_{\text{wall}}}{a}. \tag{9.2.3}$$

Without viscosity or other nonideal effects, the plasma would overshoot this point and then snap back, impelled by image currents in the wall. If the plasma were stopped at the new equilibrium state, it would tend to wander to the wall since there would be no forces to keep it centered in the uniform magnetic field. This simple example illustrates only the role of flux conservation during the nonlinear evolution. It does not consider the driving mechanism of the instability or the energetics leading to a balance of forces everywhere on the plasma. The condition for instability, $q_a < 1$, does not even appear.

Question 9.2.1
Would the result be any different if the vacuum were replaced by a force-free plasma or a perfectly conducting medium which initially carries no current?

A more detailed analysis was carried out by Rutherford, Furth, and Rosenbluth (1971) for a uniform current plasma in the large aspect ratio limit. Near the low-q marginal point, where the kink instability is almost wall stabilized

$$m - nq_a < 1 - X, \quad X \equiv (a/r_{\text{wall}})^{2m} \tag{9.2.4}$$

the amplitude of the helical equilibrium is approximately

$$\left(\frac{\xi_a}{a}\right)^2 = \frac{2(1 - X + nq_a - m)}{mX^2\left(m - 1 + 2m\dfrac{1 - 2X}{(1 - X)^2} + \dfrac{1}{\ln(r_{\text{wall}}/a)}\right)}. \tag{9.2.5}$$

Unfortunately, near the high-q (Kruskal-Shafranov) marginal point

$$0 < m - nq_a \tag{9.2.6}$$

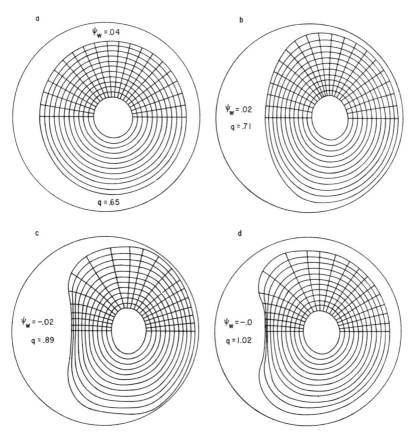

Fig. 9.2
Saturation of an $m=1$ kink mode in a large aspect ratio cylindrical plasma with shear (parabolic $J_z(r)$.) From M. N. Rosenbluth, D. A. Monticello, H. R. Strauss, and R. B. White, *Phys. Fluids*, *19*, 1987 (1976).

the amplitude of the helical equilibrium is very sensitive to the current profile within the plasma and there appears to be no reliable analytic estimate.

A computer study of free-boundary kink instabilities has been carried out by Rosenbluth, Monticello, Strauss, and White (1976). They use the helically symmetric, zero beta, large aspect ratio model described in the last section. With a parabolic current profile for the initial cylindrical equilibrium, they find that the $m=1$ kink mode shown in fig. 9.2 produces a moderate to large deformation of the plasma surface even when the wall is close to the plasma. In each frame of this illustration, a cross section of the helical equilibrium representing a state of minimum energy is shown as a function of the q-value at the edge of the plasma. Near the high-q marginal point (frame d at the lower left in fig. 9.2), there is the remnant of the "bubble" formation suggested by Kadomtsev and Pogutse (1973) where a helical filament of vacuum magnetic field tries to penetrate into the center of the plasma. The point of the bubble theory is that it is a state of lower energy to have at least part of the vacuum region on the inside of the plasma. However, this interchange is strongly suppressed by shear.

Question 9.2.2 (Kadomtsev and Pogutse (1973))
What is the change in magnetic energy when a filament of vacuum field penetrates to the center of the plasma. Assume that the plasma is incompressible and that part of the poloidal flux is carried in with the filament. When the filament reaches the center of the plasma it forms a thin straight cylinder with flux $\pi r_0^2 B_{z0}$.

As the q-value is lowered (frames c and b in fig. 9.2), the instability shifts the plasma closer to the wall with a more simple helical distortion. The low-q marginal point is shown in frame a where image currents in the wall stabilize the mode.

Cross sections of the helical equilibria resulting from an $m=2$ kink mode are shown in fig. 9.3. It can be seen that there is hardly any distortion of the plasma surface at all. This is a good example of an instability which was once thought to be dangerous because of its large growth rate and broad spatial extent compared to internal interchange modes and which now appears to be relatively harmless by itself.

It has been the experience of researchers working with high beta models and shorter longitudinal wavelengths that far from the marginal point these kink instabilities tend to drive the plasma hard into the wall where the plasma splashes like a liquid. It is not yet known what

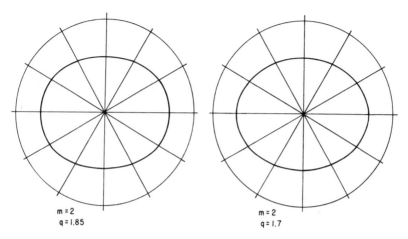

Fig. 9.3
Saturation of an $m=2$ mode in a cylinder with parabolic current. From M. N. Rosenbluth, D. A. Monticello, H. R. Strauss, and R. B. White, *Phys. Fluids*, *19*, 1987 (1976).

conditions of beta, wavelength, and viscosity are needed to produce the small deformations illustrated in figs. 9.2 and 9.3.

9.3 FIXED-BOUNDARY INSTABILITIES

Internal, fixed-boundary instabilities are characterized by velocity vortex cells rolling off each other and helically twisted down the plasma column. The $m=1$ instability, for example, consists of two vortex cells which produce a nearly uniform flow across the center of the plasma and a thinner region of return flow near the $q=1$ mode rational surface. For each $m \geqslant 2$ instability there is a ring of $2m$ vortex cells with their centers located near the $q=m$ mode rational surface. In each case the perturbed magnetic field also consists of $2m$ vortex cells corresponding to current filaments which are out of phase with the velocity vortex cells. At low aspect ratio there is also a longitudinal magnetic field perturbation running opposite to each perturbed current filament. These features in a straight cylinder were described and illustrated in section 6.5. The structures become more mixed in toroidal and strongly noncircular geometries.

Question 9.3.1
Why is most of the return flow across the cross section of the plasma and not along the length of the plasma column?

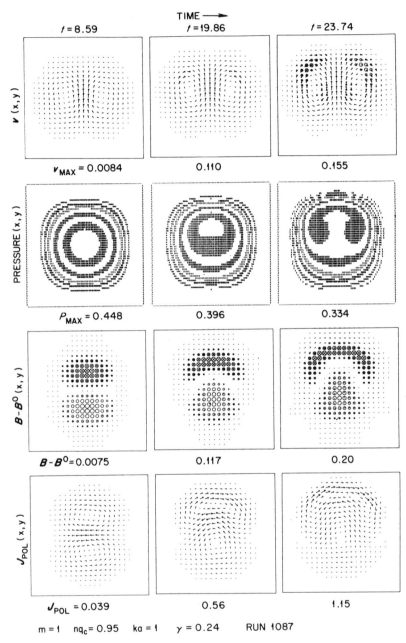

Fig. 9.4
Nonlinear evolution of velocity, pressure, perturbed magnetic field, and poloidal current for an internal $m=1$ mode near the marginal point in cylindrical geometry with $p'(\psi) \sim \psi$. From G. Bateman, H. R. Hicks, J. W. Wooten, ORNL/TM-5796 (Oak Ridge, 1977) to appear in *Nuclear Fusion*.

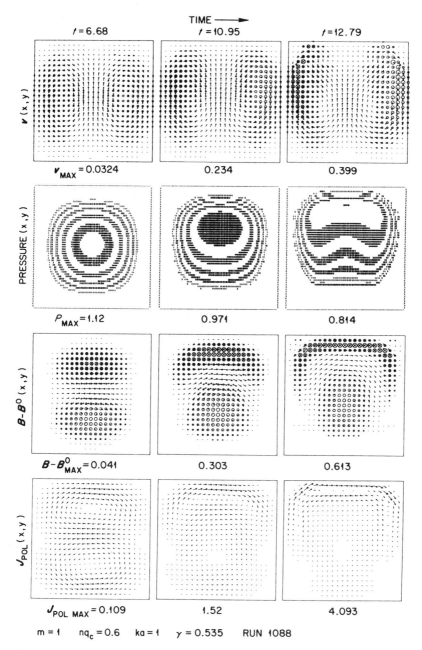

Fig. 9.5
Nonlinear evolution of an internal $m=1$ instability far from the marginal point. From G. Bateman, H. R. Hicks, and J. W. Wooten, ORNL/TM-5796.

The nonlinear evolution of each of these instabilities is dominated by convection around these vortex cells. The velocity field established during the linear phase of the instability continues to grow long into the nonlinear phase without appreciably changing structure or wandering. After the pressure, density, magnetic field and current density have been convected away from the center of the plasma and around to form an annulus of hot plasma surrounding the relatively cooler plasma drawn in from the edge, the strength of the velocity field saturates at a few tenths of the Alfvén velocity and the vortex pattern breaks up. This process, up to the point of saturation, is illustrated in figures 9.4, 9.5, and 9.6 taken from a computer simulation of nonlinear instabilities in a short cylindrical plasma by Bateman, Hicks, and Wooten (1977).

Note that the effect of an instability depends very much on its spatial extent. For example, the localized $m=1$ instability shown in fig. 9.4, where the q-value dips below one only near the center of the plasma, produces a strong mixing of the plasma near the center, leaving the edge of the plasma essentially untouched. In the next chapter we will see that resistivity plays a crucial role in completing this mixing process by allowing magnetic field lines to reconnect and rearrange themselves after they have been bent by convection. The much stronger $m=1$ instability in fig. 9.5, where the q-value is less than one through most of the plasma, has the effect of driving the plasma into the wall. Motion pictures of this process show subsequent splashing and chaotic motion.

Each of these instabilities creates a pocket of paramagnetism and diminished poloidal field near the center of the plasma, surrounded by a ring of reduced toroidal field and enhanced poloidal field. Longitudinal field reversal at the edge of the plasma has been observed in computer simulations by Sykes and Wesson (1976) when the instability is sufficiently strong.

Question 9.3.2
The total magnetic field is observed to remain helical around the geometric center of the plasma during the nonlinear evolution of the fixed-boundary $m=1$ instability. How can the $m=1$ convection carry a helical field into a similar helical field?

Computational results demonstrating the importance of large-scale convection around essentially fixed velocity vortex cells have been obtained for a variety of plasma conditions. Gerlach and Zueva (1974) have followed the convection of an $m=2$ MHD mode with resistivity

Nonlinear Instability Theory

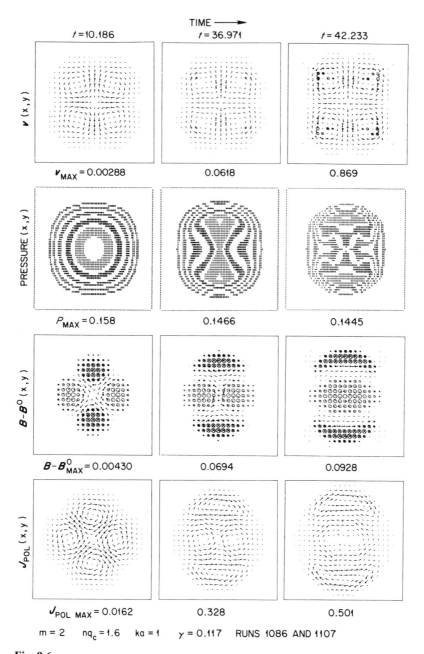

Fig. 9.6
Nonlinear evolution of an internal $m=2$ instability. From G. Bateman, H. R. Hicks, and J. W. Wooten, ORNL/TM-5796.

added for several cycles of the convective motion to observe the formation and mixing of a complex filamentary structure. Bateman, Hicks, and Wooten (1975) and Strauss (1975) have observed the effects of convection on cylindrical plasma with elongated cross sections. Strauss used the low beta equations developed by Rosenbluth, and he demonstrated the same convective phenomena with both a Lagrangian and an Eulerian computer code. Studies in toroidal geometry have revealed essentially the same process.

In the limit of low beta, large aspect ratio, and high shear, there is an analytic model for the ideal $m=1$ mode in a circular cylinder by Rosenbluth, Dagazian, and Rutherford (1973). They find a good analytic approximation to the solution of the nonlinear equations in a thin boundary layer around the $q=1$ mode rational surface, and they match this solution onto the linear solution everywhere else, using a procedure similar to that in section 6.5. A saturation amplitude is found for the displacement of the center of the plasma column

$$\xi_a = \frac{13}{r_s^3 [(\mathbf{k} \cdot \mathbf{B})'_{r_s}]^2} \int^{r_s} dr(-4\pi g) \tag{9.3.1}$$

where r_s is the radius of the $q=1$ surface, the driving term g is given by (6.4.7), and $\mathbf{k} \cdot \mathbf{B} = B_\theta(r)/r - kB_z(r)$. It is difficult for computational methods to approximate the conditions under which this analytic treatment applies.

Question 9.3.3
What is the limiting displacement for an $m=1$ mode in an equilibrium with parabolic current profile $J_z = J_{z0}(1 - r^2/a^2)$, with r_s bounded away from the origin? When does the displacement become smaller than r_s?

9.4 REFERENCES

A survey of recent nonlinear MHD computer simulations can be found in the review article by

J. A. Wesson, *Computer Physics Communications*, 12, 53–65 (1976).

The computation of S. J. Roberts and A. Curtis was reported by

W. B. Thompson et al., *Geneva Conference*, 32, 65–71 (1958).

Other references on computer simulation cited in the text are

G. Van Hoven and M. A. Cross, *Phys. Rev.*, A7, 1347–1352 (1973).

J. U. Brackbill, *Methods in Computational Physics*, 16, 1–41 (1976).

M. N. Rosenbluth, D. A. Monticello, H. R. Strauss, and R. B. White *Phys. Fluids*, *19*, 1987–1996 (1976).

H. R. Strauss, *Phys. Fluids*, *19*, 134–140 (1976).

N. E. Gerlach, N. M. Zueva et al., Institute of Applied Mathematics (Moscow), preprint numbers 89, 91 (1974), 74, 95, 96, 111 (1975).

G. Bateman, H. R. Hicks, J. W. Wooten, ORNL/TM-5796 (1977), Proceedings of the 3rd Topical Conference on Pulsed High Beta Plasmas, Culham, England (1975).

A. Sykes and J. A. Wesson, *Phys. Rev. Letters*, *37*, 140–144 (1976).

On bifurcation theory, see

K. O. Friedrichs, *Rev. Mod. Phys.*, *32*, 889–897 (1960).

T. Yeh, *Phys. Fluids*, *16*, 516–528 (1973).

P. H. Rutherford, H. P. Furth, M. N. Rosenbluth, IAEA Madison Conference *2*, 553–570 (1971).

B. B. Kadomtsev, *Reviews of Plasma Physics*, *2*, 188–190 (1963).

B. B. Kadomtsev and O. P. Pogutse, Sixth European Conf. on Controlled Fusion and Plasma Physics, Moscow, *1*, 59–62 (1973); *Sov. Phys.*—JETP, *39*, 1012–1016 (1974).

On the related subject of singular perturbation theory, see

M. N. Rosenbluth, R. Y. Dagazian, P. H. Rutherford, *Phys. Fluids*, *16*, 1894–1902 (1973).

P. H. Rutherford, *Phys. Fluids*, *16*, 1903–1908 (1973).

On convection cells:

S. Chandrasekhar, *Hydrodynamic and Hydromagnetic Stability* (Oxford: Clarendon Press, 1961).

H. Wobig, *Plasma Physics*, *14*, 403–416 (1972).

E. K. Maschke and R. B. Paris, report EUR-CEA-FC-721 (1974); IAEA Tokyo Conf., *1*, 531–539 (1974).

R. Y. Dagazian and R. B. Paris, *Phys. Fluids*, *20*, 917–927 (1977).

W. L. Harries, Phys. Fluids, *13*, 1751–1761 (1970).

On fully developed MHD turbulence, see

B. B. Kadomtsev and O. P. Pogutse, *Reviews of Plasma Physics*, *5*, 249–400 (1967).

R. H. Kraichnan, *Phys. Fluids*, *10*, 1417–1423 (1967).

D. Fyfe and D. Montgomery, *J. Plasma Phys.*, *16*, 181–191 (1976).

D. Fyfe, D. Montgomery, and G. Joyce, *J. Plasma Phys.*, *17*, 317–335, 369–398 (1977).

Additional work has been done on the nonlinear evolution of resistive tearing modes which will be discussed in the next chapter.

10
Resistive Instabilities

Up to this point we have considered only ideal MHD instabilities. It might be expected that the addition of resistivity or viscosity or heat conductivity to the ideal equations would only reduce the growth rate of instabilities because resistivity would dissipate away electrical currents, viscosity would damp out any sheared velocities, and heat conductivity would diminish local thermal gradients which could provide a source of free energy. However, it is important to realize that the addition of dissipation can produce new instabilities by removing constraints from the ideal equations and, thereby, making states of lower potential energy accessible to the plasma. This relaxation of constraints can also make existing instabilities grow faster. A new kind of instability called the resistive tearing mode, which appears when a small amount of resistivity is added to the ideal MHD equations, will be the main subject of this chapter.

The addition of resistivity allows magnetic field lines to break and reconnect so that they are no longer frozen into the fluid as described in section 2.2. This has a profound effect on the MHD model. With resistivity, the magnetic field tends to break up into a number of thin filaments called magnetic islands which thread their way through the plasma. Since heat flows rapidly along field lines, one of the direct effects of this island structure is to enhance transport across the plasma. The concept of magnetic islands will be examined in the next section.

Resistive Instabilities 191

The growth rate of resistive instabilities is determined by how fast the magnetic field lines can be broken, viewed as a kind of throttling process, balanced against the driving forces and inertial effects. It is an important feature of these instabilities that the field lines need to be broken only in a thin boundary layer within the plasma. Since the dissipation process is governed by a diffusion equation, the time scale for the throttling process is determined by the width of the boundary layer squared divided by the diffusion coefficient. Dissipation effects typically add terms with higher order spatial derivatives to the ideal equations and this naturally leads to a singular perturbation analysis of boundary layers. If the boundary layer thickness were comparable to the radius of the plasma, the result would be resistive diffusion which proceeds on a very slow time scale compared to most MHD instabilities. If the boundary layer were too thin, inertial effects would limit the growth rate of the instability. When all the effects are balanced, it turns out that the resistive instabilities grow on a time scale much slower than the Alfvén transit time scale and much faster that the typical diffusion time scale. In section 10.2, a simple estimate will be made for the boundary layer thickness and growth rate of the tearing mode, which tends to break the plasma into magnetic islands. In the same section the nonlinear development of $m \geq 2$ magnetic islands will be outlined.

The $m = 1$ resistive kink instability, which is a special case of a tearing mode, will be treated in section 10.3. This mode is believed to cause the sawtooth oscillations observed in the soft X rays emitted by tokamaks. Computational studies of its nonlinear development will also be reviewed.

The resistive interchange mode driven by an inverted pressure gradient in a curved magnetic field, sometimes called the resistive g-mode when it is driven by an inverted density gradient in a gravitational field, will be considered in section 10.4. Finally, a stability criterion derived by Glasser, Greene, and Johnson (1975) that includes both the tearing and the interchange modes in toroidal geometry will be examined at the end of section 10.4.

A full review of the literature is far beyond the scope of this chapter. Instead, the simplest possible estimates and derivations will be given to elucidate the basic physics of resistive instabilities. Many people find the literature in this field hard to read, partly because of the notation used, partly because at least a fourth order system of differential equations must be used—even in the simplest geometry—and partly because several effects must be balanced simultaneously to derive

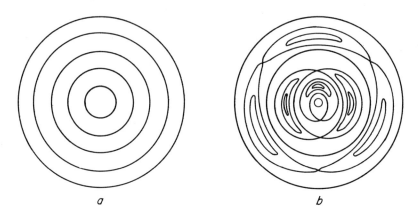

Fig. 10.1
A schematic cross section of (a) simply nested flux surfaces compared to (b) flux surfaces with $m = 1$, 2, and 3 island structures.

the ordering procedures that make the approximations hang together. Here, the notation will be consistent with the rest of the book, the assumptions will be isolated, and the derivations will be presented with as little algebra as possible.

10.1 MAGNETIC ISLANDS

Up to this point, we have been considering that particular class of MHD equilibria composed of simply nested flux surfaces around a single magnetic axis, as shown in fig. 10.1a. We must now consider a wider class of magnetic field configurations which have helical magnetic island structures located at some of the rational surfaces, as in fig. 10.1b. A *magnetic island* is a filament of plasma with its own set of nested flux surfaces surrounding its own local magnetic axis. Each island twists helically around the torus following the closed field line that forms its magnetic axis. The whole structure of each island closes upon itself after going the long way around the torus one or more times, depending upon the rational order of its magnetic axis.

Magnetic islands can be traced out by following a field line within the island many times around the torus until the surface covered by that field line becomes clear. Normally, the cross-sectional shape of the island is determined by interpolating a curve between the points where a field line intersects a cross section of the torus. For analytic work, however, it is easier to determine the shape of magnetic islands from the helical flux function, which will be illustrated here.

The first step is to consider a cylindrical or toroidal magnetic field configuration with simply nested flux surfaces and finite shear ($q(\psi)$ not uniform). Consider a mode rational surface where $q = m/n$ within this configuration. Then construct another magnetic field with exactly the same flux surfaces but zero shear, so that $q = m/n$ everywhere within the plasma. Subtract this artificial zero shear magnetic field from the original plasma magnetic field to define

$$\mathbf{B}_*(\mathbf{x}) \equiv \mathbf{B}(\mathbf{x}) - \mathbf{B}_{q=m/n}(\mathbf{x}). \tag{10.1.1}$$

For example, in a circular cylinder this $\mathbf{B}_*$ field would be

$$\mathbf{B}_* = \mathbf{B} - \frac{r}{r_s} B_\theta(r_s)\hat{\boldsymbol{\theta}} - B_z(r_s)\hat{\mathbf{z}} \tag{10.1.2}$$

where r_s is the radius of the mode rational surface on which $q = 2\pi r_s B_z(r_s)/LB_\theta(r_s)$ and L is the length of the cylinder under consideration. The poloidal component of this $\mathbf{B}_*$ field changes sign across the mode rational surface by going through zero at the surface. Keep in mind that in systems with curvature there is a current density associated with the zero shear field (a uniform current for the circular cylinder example given here) that has been subtracted from the original configuration when defining $\mathbf{B}_*$.

In order to visualize how magnetic islands are formed, imagine that the cylindrical or toroidal plasma is cut open and laid out flat, as illustrated in fig. 10.2a–c. This process is equivalent to the transformation from Cartesian coordinates to a flux coordinate system in which the field lines are straight such as Hamada coordinates, as discussed in section 7.1. Then the transformation from $\mathbf{B}$ to $\mathbf{B}_*$ is equivalent to changing our orientation so that we are looking straight down the field lines at the mode rational surface, as shown in fig. 10.2d. From this point of view, the poloidal component of $\mathbf{B}_*$ runs from left to right above the mode rational surface and from right to left below it.

Now let there be a small radial magnetic field perturbation with sinusoidal variation along the poloidal angle θ, as illustrated in fig. 10.2e. In the region where this radial field is positive, a field line close below the mode rational surface will move up through the rational surface and out to where the poloidal component of $\mathbf{B}_*$ is positive. There the field line will move to the right until it gets to a region where the radial field is negative. This takes the field line back down through the rational surface where it moves to the left and once again gets to a region where the radial field is positive. This cyclic orbit of the field lines near the rational surface forms the magnetic islands shown in fig. 10.2e. Far above the rational surface the field lines continue to

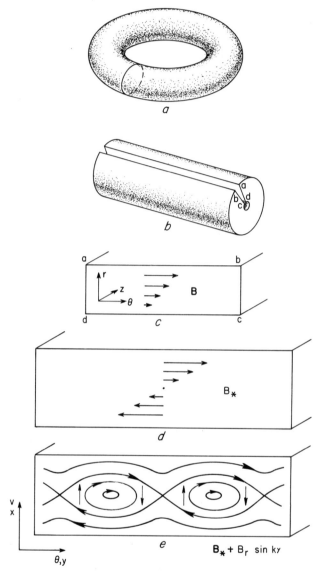

Fig. 10.2
Toroidal plasma, cut open and laid out flat, with transformation to B_* and a radial magnetic field perturbation to show the formation of magnetic islands.

Resistive Instabilities

move from left to right, undulating under the influence of the radial field but never staying in any one part of the radial field long enough to be forced through the rational surface. Therefore, a last closed magnetic surface forms a separatrix between the surfaces that close around the islands and the open surfaces above and below. Of course all the magnetic field orbits shown in fig. 10.2e are projections of the field lines on a cross-sectional plane. In general, the actual field lines have a component into or out of the paper.

In order to see why resistivity is needed to form an island or to change it in time, imagine a contour that runs the long way around the toroid along the field line at the center of an island and returns along the field line at an adjacent x-point. Since the rational surface containing these field lines closes on itself around the torus, the connecting links of the contour can be made to cancel. For the island to be changing in time, the radial magnetic field through the rational surface must be changing in time. For this to happen, Faraday's law indicates that there must be an electric field around the contour. However, the contour is parallel to the field lines, except for the parts that cancel. An electric field parallel to a magnetic field line is forbidden in the ideal MHD model where $\mathbf{E} = -\mathbf{v} \times \mathbf{B}$. However, with finite resistivity, a parallel electric field merely produces a current that flows along the magnetic island.

It is a simple matter to derive the width of a magnetic island given the $\mathbf{B}_*$ field and the radial field perturbation. First consider the derivation in a plane slab geometry, where x corresponds to the coordinate along the minor radius of the toroid and y corresponds to the poloidal angle. Let the mode rational surface be at $x=0$ and use a series expansion for B_* in the neighborhood of this surface

$$B_{*y}(x) = B'_{*y} x + \cdots. \tag{10.1.3}$$

Consider only one harmonic of the radial magnetic field perturbation

$$B_x^1(x, y) = B_x^1 \sin k_y y. \tag{10.1.4}$$

The total magnetic field can be described by a flux function which ensures $\nabla \cdot \mathbf{B} = 0$

$$B_x = -\frac{\partial \psi}{\partial y}, \qquad B_y = \frac{\partial \psi}{\partial x} \tag{10.1.5}$$

$$\psi = \psi_0 + \tfrac{1}{2} B'_{*y} x^2 + \cdots + (B_x^1/k_y)\cos k_y y. \tag{10.1.6}$$

Resistive Instabilities

When calculating the current, it should be noted that B_{*y} is perturbed as well as B_{*x}.

The level contours of this flux function appear as in fig. 10.2e. At the x-point of the separatrix, the flux function assumes its minimum value along the $x=0$ line at $k_y y = -\pi/2$

$$\psi_s = \psi_0 - B_x^1/k_y. \tag{10.1.7}$$

Let W stand for the maximum width of the separatrix. The separatrix is widest at the point $(x = \pm W/2, k_y y = \pi/2)$, along the vertical line passing through the o-point of the magnetic island at $x=0$

$$\psi_s = \psi_0 + \tfrac{1}{2} B'_{*y} \left(\frac{W}{2}\right)^2 + \cdots + B_x^1/k_y. \tag{10.1.8}$$

Eliminating $\psi_s - \psi_0$ from (10.1.7) and (10.1.8) yields an expression for the island width in a plane slab

$$W = 4(B_x^1/k_y B'_{*y})^{1/2}. \tag{10.1.9}$$

This derivation can be easily extended to any axisymmetric toroidal configuration. Choose V, θ, φ as the coordinate system, where V is the volume of each flux surface, θ is any choice of poloidal angle around the flux surfaces, and φ is the toroidal angle. The $\mathbf{B}_*$ field, defined by (10.1.1) for this system of simply nested flux surfaces, can be described by a helical flux function ψ_*

$$\mathbf{B}_* = \nabla \psi_* \times \nabla \varphi + B_{*\varphi} \hat{\varphi}. \tag{10.1.10}$$

Now consider one harmonic of the radial field perturbation, which can be included in the flux function

$$\psi = \psi_*(V) + \psi_r \cos(m\theta - n\varphi). \tag{10.1.11}$$

Use a series expansion for ψ_* in the neighborhood of the mode rational surface at $V = V_0$, on which $B_* = 0$,

$$\psi = \psi_{*0} + \tfrac{1}{2} \ddot{\psi}_*(V_0)(V - V_0)^2 + \cdots + \psi_r \cos(m\theta - n\varphi). \tag{10.1.12}$$

Following the argument given above, the volume width of the magnetic island is given by

$$|V_s - V_0| = 2(\psi_r/\ddot{\psi}_*)^{1/2}. \tag{10.1.13}$$

Here, V_s is to be interpreted as that volume of the unperturbed flux surface which passes through the separatrix at its widest point. For the special case of a circular cylinder, we have

$$V_s - V_0 = \pi L(r_s - r_0)^2 \simeq \frac{W}{2}\pi L 2 r_s$$

$$B_{*\theta}\hat{\theta} + B_r \hat{r} = 4\pi^2 r \dot{\psi}_* \hat{\theta} + \cdots + \frac{2\pi m}{rL}\psi_r \sin(m\theta - n\varphi)\hat{r}$$

$$W \simeq 4r\left(\frac{B_r^1}{mB_\theta}\left|\frac{q}{rq'(r)}\right|\right)^{1/2} \tag{10.1.14}$$

where r_s is the radius of the mode rational surface, and m is the number of islands the short way around the cylinder. For wider islands the effects of radial dependence of B_r^1 and shear over the island width must be taken into account.

Question 10.1.1 (J.M. Finn (1975))
Suppose many harmonics of the radial field perturbation are given, corresponding to many rational surfaces within the plasma. Is there more chance for island overlap with high shear or low shear? Island overlap is believed to enhance greatly transport across the plasma.

Question 10.1.2
What is the current density perturbation within a magnetic island? Is it possible to have islands in a vacuum magnetic field?

Question 10.1.3
Within a magnetic island, how many times do we have to follow a field line around the torus before it moves from the inside of the island to the outside? What is the q-value relative to the island's own magnetic axis?

Question 10.1.4
Do all the field lines within a magnetic island have the same q-value relative to the main magnetic axis of the torus? If so, is this q-value rational?

Question 10.1.5
Which has less energy stored in the magnetic field, the slab without islands as in fig. 10.2d or the slab with islands as in fig. 10.2e? Assume that B_x^1 decreases away from $x=0$ until it is zero at $x=w$. Let the poloidal field perturbation B_y^1 be determined so that the flux between the center of the island and the wall at $x=w$ is fixed.

10.2 GROWTH OF THE RESISTIVE TEARING MODE

A *tearing mode* is the tendency for a plasma to break up into magnetic islands in order to reduce the magnetic energy in the regions away from the islands. As the islands grow, flux surfaces are pulled in from above and below to weld together at the x-points and then tear apart to form the closed surfaces within the islands. The structure of the radial magnetic field perturbation that produces these islands must be determined by solving the ideal MHD equations as a boundary value problem in the regions away from the islands. All the rest of the action occurs in a thin boundary layer which surrounds the islands. This boundary layer is where the growth of the island width is determined by balancing the driving forces against the resistive diffusion and inertial effects.

A very simplified derivation will be given in this section to estimate the growth rate of a resistive tearing mode in a plane slab geometry. This derivation applies only to $m \geqslant 2$ tearing modes if the slab is considered to be an approximation to a cylinder or a torus as illustrated in fig. 10.2a–c. The $m = 1$ mode is a special case that will be treated in the next section. For the $m \geqslant 2$ modes, it will be shown that the growth rate remains exponential only as long as the island width is much smaller than the boundary layer thickness. Then, because of nonlinear effects as the island width becomes comparable to the boundary layer thickness, the island width starts to grow as a linear function of time. Finally, the island width saturates at some fraction of the plasma width, depending upon the form of the resistivity profile.

For an estimate of the linear tearing mode growth rate, consider a plane slab configuration and only a simplified subset of the resistive MHD equations. The equilibrium is characterized by the poloidal field B_* defined by (10.1.1) as described in the last section. As before, $B_* = 0$ at $x = 0$, where x corresponds to the radial coordinate perpendicular to the equilibrium flux surfaces.

Only the radial component of Faraday's law will be needed

$$\frac{\partial}{\partial t} B_x^1 = -\frac{\partial}{\partial y} E_z^1. \tag{10.2.1}$$

The z-derivative vanishes in all the equations because everything is assumed to be uniform along the length of the magnetic islands. The electric field along the z-direction is given by

$$E_z^1 = -v_x^1 B_* + \eta^0 J_z^1 \tag{10.2.2}$$

Resistive Instabilities

where the resistivity η^0 is assumed to be uniform here (it will be important only in a thin boundary layer anyway). Only one driving force term will be considered in this simplified model

$$\rho^0 \frac{\partial}{\partial t} v_x^1 = -J_z^1 B_* \tag{10.2.3}$$

and all motion will be assumed to be incompressible

$$\nabla \cdot \mathbf{v}^1 = 0. \tag{10.2.4}$$

Question 10.2.1
Are (10.2.4) and (10.2.3) compatible?

Finally, it will be assumed that there is no velocity or perturbed magnetic field along the islands

$$v_z^1 = 0, \qquad B_z^1 = 0. \tag{10.2.5}$$

In a more detailed analysis, most of these can be shown to be good approximations. Here they will just be taken as a simplified model.

Let the perturbed variables B_x^1 and v_y^1 vary like $\sin ky \, e^{\gamma t}$ while E_z^1, J_z^1, v_x^1 and B_y^1 vary like $\cos ky \, e^{\gamma t}$, as shown in fig. 10.3. From $\nabla \cdot \mathbf{B} = 0$ it follows that

$$\frac{\partial}{\partial x} B_x^1 = k B_y^1 \tag{10.2.6}$$

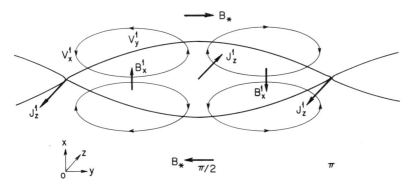

Fig. 10.3
The magnetic field, longitudinal current density, and flow pattern of a tearing mode are shown together with the separatrix of the induced magnetic island.

and

$$J_z^1 = \frac{\partial B_y^1}{\partial x} - \frac{\partial B_x^1}{\partial y} = \frac{1}{k}\frac{\partial^2 B_x^1}{\partial x^2} - kB_x^1. \tag{10.2.7}$$

From (10.2.1), (10.2.2), and (10.2.7) we have

$$\frac{\partial B_x^1}{\partial t} = v_x^1 k B_* - \eta^0 \left(\frac{\partial^2 B_x^1}{\partial x^2} - k^2 B_x^1 \right). \tag{10.2.8}$$

If the resistivity η^0 is small enough, it can be neglected everywhere except near $x = 0$ where B_* vanishes. At $x = 0$, we have a singular differential equation in which a finite $\partial B_x^1/\partial t$ term must be balanced by a large value of $\partial^2 B_x^1/\partial x^2$ times the small value of η^0.

The key to this resistive instability calculation is that the quantities $\partial^2 B_x^1/\partial x^2$ and J_z^1 are large only in a thin boundary layer around the mode rational surface. When viewed on the scale of the plasma radius, the radial derivative of B_x^1 will appear to have a discontinuity at the radius of the mode rational surface

$$\Delta' \equiv (B_x^{1\prime}(r_s + \varepsilon) - B_x^{1\prime}(r_s - \varepsilon))/B_x^1(r_s) \tag{10.2.9}$$

where the radius $r = r_s$ in the cylinder corresponds to $x = 0$ in the plane slab. This approximation gives the scale size for $\partial^2 B_x^1/\partial x^2$ within the boundary layer

$$\frac{\partial^2 B_x^1}{\partial x^2} \simeq \Delta' B_x^1/\varepsilon \tag{10.2.10}$$

where ε is the width of the boundary layer. Using (10.2.10) and (10.2.7) yields an approximation for the longitudinal current density within the tearing mode magnetic island

$$J_z^1 \simeq \Delta' B_x^1/k\varepsilon. \tag{10.2.11}$$

The parameter Δ' appears throughout the literature on linear resistive tearing modes. It is the essential link between the boundary layer and the rest of the plasma. It is determined by computing the radial magnetic field perturbation in the regions outside the boundary layer using the ideal MHD equations subject to boundary conditions at the wall and at the main magnetic axis of the plasma. The radial field computed in this way must be continuous at the rational surface while the discontinuity in its first derivative determines Δ'. In this section, Δ' will be taken as a given quantity. It has been computed in a circular cylinder with several profiles by Furth, Rutherford, and Selberg (1973).

Glasser, Furth, and Rutherford (1977) have developed a computer program to determine Δ' for any experimentally measured profile. In the plane slab the full eigenvalue problem has been solved in the classic paper by Furth, Killeen, and Rosenbluth (1963).

In order to estimate the thickness of the boundary layer and the growth rate of the corresponding tearing mode, balance the magnitudes of the three terms in Faraday's law (10.2.1) and (10.2.2) within the boundary layer

$$\gamma B_x^1 \simeq v_x^1 k B_* \simeq \eta^0 k J_z^1. \qquad (10.2.12)$$

From the first and the last terms in this equation and from (10.2.11) it follows that

$$\gamma \simeq \eta^0 \Delta'/\varepsilon. \qquad (10.2.13)$$

The width of the boundary layer ε must now be determined in order to estimate the growth rate.

The equation for ε comes from balancing the inertial effects against the driving forces. In order to estimate the kinetic energy in the boundary layer, observe that v_y is much larger than v_x, as illustrated in fig. 10.3. This follows from the incompressibility condition $\mathbf{V} \cdot \mathbf{v} = 0$ and the fact that the poloidal wavelength is assumed to be much longer than the width of the boundary layer $k\varepsilon \ll 1$

$$k v_y^1 \simeq v_x^1/\varepsilon. \qquad (10.2.14)$$

Then the rate of change of kinetic energy is

$$\gamma \rho \tfrac{1}{2}[(v_x^1)^2 + (v_y^1)^2] \simeq \gamma \rho \tfrac{1}{2}(v_x^1)^2/(k\varepsilon)^2. \qquad (10.2.15)$$

This is to be balanced against the rate at which work is done on the fluid, using the driving force from (10.2.3)

$$\mathbf{v}^1 \cdot \mathbf{F} \simeq v_x^1 B_* J_z^1 \qquad (10.2.16)$$

to obtain

$$\gamma \rho v_x^1 \simeq 2\Delta' B_* k\varepsilon B_x^1. \qquad (10.2.17)$$

For the purposes of making these estimates, the typical value of B_* within the boundary layer is taken to be

$$B_* \simeq B_*' \varepsilon. \qquad (10.2.18)$$

The final step is to substitute the expression for v_x^1 from (10.2.17) into (10.2.13) and then evaluate ε and γ. The result for the width of the boundary layer is

$$\varepsilon \simeq \left(\frac{\rho \eta^2 \Delta'}{2(kB'_*)^2}\right)^{1/5} \simeq S^{-2/5}(\Delta' a)^{1/5}\left(\frac{a}{R}n\frac{aq'}{q}\right)^{\pm 2/5} \quad (10.2.19)$$

The growth rate then follows directly from (10.2.13)

$$\gamma \simeq .55(\Delta')^{4/5}\left(\frac{\eta^3(kB'_*)^2}{\rho}\right)^{1/5} \quad (10.2.20)$$

where the factor .55 comes from a more exact treatment. This growth rate for the resistive tearing mode may be written

$$\gamma \simeq .55\,\tau_R^{-3/5}\,\tau_A^{-2/5}(\Delta' a)^{4/5}\left(\frac{a}{R}n\frac{aq'}{q}\right)^{2/5} \quad (10.2.21)$$

where n is the toroidal mode number and

$$\tau_R \equiv a^2 \mu/\eta \quad (10.2.22)$$

is the resistive skin time across the plasma radius and

$$\tau_A \equiv a/V_A, \quad V_A \equiv B_{z0}/(\mu\rho)^{1/2} \quad (10.2.23)$$

is the Alfvén transit time across the plasma radius. In this form the growth rate can be seen to be 2/5 Alfvénic and 3/5 dissipative—part way between the MHD and the transport time scales. The parameter

$$S \equiv \tau_R/\tau_A \quad (10.2.24)$$

is known as the magnetic Reynolds number. Since S is typically on the order of 10^6 in hot tokamak plasmas, it is clear that the Alfvén and the resistive time scales are well separated and the boundary layer thickness is very small compared to the radius of the plasma, in spite of its 2/5 power dependence on S.

Question 10.2.2
The condition

$$\Delta' > 0 \quad (10.2.25)$$

is a necessary prerequisite for a tearing mode instability. Why?

Question 10.2.3
How would an externally applied helical field alter the growth rate of its resonant tearing mode? Would the externally applied field increase or decrease Δ'?

Resistive Instabilities

Question 10.2.4
For hot tokamak plasmas ($T_e \sim T_i \sim 1\text{--}10$ keV, $n_e \sim 10^{14}$ cm^{-3}, $B \sim 5$ Tesla, $a \sim 20\text{--}50$ cm, $R \sim 100\text{--}200$ cm), how thick is the resistive boundary layer compared to the ion Larmor radius? How large is the growth rate compared to the collision frequency or the diamagnetic drift frequency? These are among the estimates that need to be made to determine the limits of applicability for resistive MHD theory.

Now consider the nonlinear development of a tearing mode. Rutherford (1973) showed that when the magnetic island width grows to be as large as the boundary layer width, a new force becomes important and the growth rate of the instability is greatly reduced. Instead of growing like an exponential function in time, the island width starts to grow like a linear function in time. This will be demonstrated with simple estimates in the next few paragraphs.

Referring once again to fig. 10.3, it can be seen that the velocity field can drive a second order contribution $v_y^1 B_x^1$ to the electric field which, in turn, can drive a second order current density along the magnetic islands

$$J_z^{(2)} \simeq v_y^1 B_x^1 / \eta^0. \qquad (10.2.26)$$

This current has a $\sin^2 ky$ spatial dependence, and it produces a new $J_z^{(2)} B_x^1$ force which opposes the v_y^1 flow everywhere. To estimate the effect of this new force, balance the rate at which work is being done by the new force against the old rate (10.2.16)

$$v_y^1 J_z^{(2)} B_x^1 \simeq v_x^1 J_z^1 B_*. \qquad (10.2.27)$$

Using (10.2.11) to (10.2.18), it is a matter of direct substitution to show that these rates balance when

$$\varepsilon \simeq (B_x^1 / k B_*')^{1/2}. \qquad (10.2.28)$$

The expression on the right in this equation is simply the island width in the plane slab geometry (10.1.9). Therefore, (10.2.28) implies that the growth rate of the tearing mode is reduced when the island width becomes as large as the resistive boundary layer width (10.2.19).

We can now estimate the growth of the island width under these new circumstances. It can be shown by a more rigorous derivation or by computation that the magnetic island structure retains its $\sin ky$ dependence and that higher harmonics are not important for this estimate. Then (10.2.1) and (10.2.2) may be combined and approximated by

$$\frac{\partial}{\partial t} B_x^1 \simeq k\eta^0 J_z^1. \qquad (10.2.29)$$

With the same arguments that were used to derive (10.2.11), it can be shown that the longitudinal current density spread out across the island width is approximately

$$J_z^1 \simeq \Delta' B_x^1 / kW \qquad (10.2.30)$$

where W is the island width given by (10.1.9) in a plane slab or (10.1.14) in a circular cylinder. Eliminating the radial field perturbation B_x^1 in favor of the island width W, we obtain

$$\frac{d}{dt} W \simeq 1.66 \eta \Delta' / k \qquad (10.2.31)$$

where the factor 1.66 comes from a more rigorous derivation. Since η and Δ' are taken to be constants in time, (10.2.31) indicates that the island width grows as a linear function of time. The essential point in this derivation is that the boundary layer thickness is replaced by the island width.

Finally, it has been shown by White, Monticello, Rosenbluth, and Waddell (1977) that the growth of the island width saturates when the islands become wide enough according to the equation

$$\frac{d}{dt} W \simeq 1.66 \eta(r_s)[\Delta'(W) - \alpha W] k \qquad (10.2.32)$$

where r_s is the radius of the original mode rational surface, $\Delta'(W)$ is the discontinuity in the derivative of the radial magnetic field perturbation from one edge of the island to the other, and α is a numerical constant which is obtained by matching the flux and current density across the separatrix of the island. If the resistive equilibrium fields are used in a straight circular cylinder

$$\eta^0(r) J_z^0(r) = E^0 \qquad (10.2.33)$$

where E^0 is a uniform externally applied electric field, then α is given by

$$\alpha \simeq \frac{m^2}{r_s^2} - \frac{\delta}{r_s} - \frac{1.1\delta}{kB'_*}\frac{dJ_z^0}{dr} - \frac{.4}{kB'_*}\frac{d^2 J_z^0}{dr^2} \qquad (10.2.34)$$

where $\delta \equiv (B_r^1)'/B_r^1$ accounts for radial asymmetry of B_r^1 at the mode rational surface. Like Δ', the variable α must be determined by solving for the radial magnetic field perturbation in the regions outside the magnetic islands. For peaked equilibrium current profiles, it is found that the saturation width is typically a few tenths of the plasma radius.

In summary then, after a brief phase of exponential growth (10.2.21) the tearing mode island width starts to grow as a linear function of time (10.2.31) when the island width becomes comparable to the resistive boundary layer thickness and finally the island width saturates according to (10.2.32).

10.3 $M=1$ RESISTIVE TEARING MODE

It was demonstrated in section 6.4 that the internal $m=1$ kink mode is unstable whenever the q-value drops below $1 + \mathcal{O}(ka)$ in the center of a circular cylindrical plasma column, where k is the longitudinal wave number and a is the radius of the current channel. When the $q=1$ mode rational surface is in a high shear region away from the center of the plasma, the structure of the instability can be divided into two regions. Within the $q=1$ surface, each cross sectional slice of the plasma has a nearly uniform transverse velocity, with the direction of this velocity rotating through 360° along the length of the instability. Then, in a thin boundary layer around this region, there is a large return velocity along the $q=1$ surface to provide a closed path for the plasma motion. There is very little flow or perturbed magnetic field outside the $q=1$ surface. The growth rate (6.5.9) was estimated in section 6.5 by balancing the driving force of the instability within the $q=1$ surface against the inertia dominated by the flow field within the boundary layer. Under these conditions it is clear that any nonideal effects which significantly alter the flow within the boundary layer will take the place of inertia in determining the growth rate of this instability.

When there is sufficient resistivity in the boundary layer, an $m=1$ tearing mode with one magnetic island is produced with a linear growth rate of

$$\gamma = (q'(r_s)ka^2 S)^{2/3} \tau_R^{-1} \qquad (10.3.1)$$

where S and τ_R are defined by (10.2.24) and (10.2.22). Actually, the ideal growth rate (6.5.9) and the resistive growth rate (10.3.1) are two limits of a more complete dispersion relation derived by Coppi, Galvão, Pellat, Rosenbluth, and Rutherford (1976). In a low beta plasma, this resistive growth rate should probably be used even when it is somewhat smaller than the ideal growth rate (6.5.9) since Rosenbluth,

Dagazian, and Rutherford (1973) find that the ideal mode becomes nonlinearly saturated at a small amplitude while Waddell, Rosenbluth, Monticello, and White (1976) find that the resistive mode continues to grow until the center of the plasma column is completely mixed. Note that the growth rate of the resistive mode increases with shear ($\gamma \sim (q'(r_s))^{2/3}$) while the growth rate of the ideal mode decreases with shear ($\gamma \sim q'(r_s)^{-1}$). In this respect, both forms of this $m=1$ instability depend upon the local conditions at the $q=1$ surface.

Other transport terms also have a large effect on the boundary layer at the $q=1$ surface. Bussac, Edery, Pellat, and Soulé (1976) and Waddell, Laval, and Rosenbluth (1977) find that the ideal form of the $m=1$ mode is completely stabilized by finite Larmor radius effects if the ideal growth rate (6.5.9) is less than half the ion diamagnetic drift frequency

$$\omega_{*i} \equiv p_i'(r_s)/(en r_s B_z). \qquad (10.3.2)$$

Their analysis indicates, however, that the same effects only reduce the growth rate of the resistive form of the $m=1$ instability without completely stabilizing it. The reduced growth rate of this resistive mode is

$$\gamma \simeq \gamma_T^3/(\omega_{*i}\omega_{*e}) \qquad (10.3.3)$$

applicable in the limit

$$\gamma_k \tau_A \ll (ka)^2$$
$$\gamma_T \ll w_{*i}/2 \qquad (10.3.4)$$
$$\gamma_T \ll \omega_{*e}/2$$

where γ_T is the tearing mode growth rate (10.3.1), γ_k is the ideal mode growth rate (6.5.9), and

$$\omega_{*e} = (p'(r_s) + 0.71 n T_e'(r_s))/(en r_s B_z) \qquad (10.3.5)$$

is the electron diamagnetic frequency, where T_e is the electron temperature, n is the particle density, and r_s is the radius of the $q=1$ surface.

The nonlinear development of the $m=1$ resistive tearing mode has received a great deal of attention recently because of the role it is believed to play in producing sawtooth oscillations observed in the soft X-ray emission from tokamaks (see section 11.1). A speculative outline of this process by Kadomtsev (1975) has been largely borne out by the accurate computations of Waddell, Rosenbluth, Monticello, and White (1976). A reasonably complete picture of the sawtooth

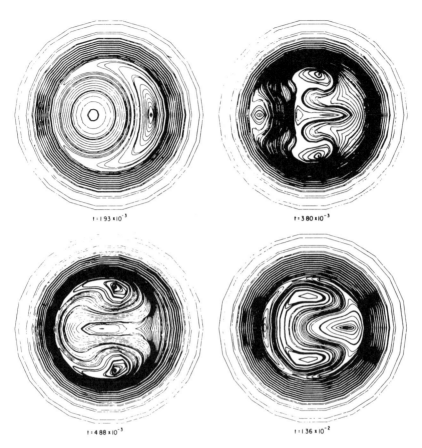

Fig. 10.4
Nonlinear evolution of the helical flux function contours for an $m = 1$ resistive tearing mode. From B. V. Waddell, M. N. Rosenbluth, D. A. Monticello, and R. B. White, *Nuclear Fusion*, *16*, 528 (1976).

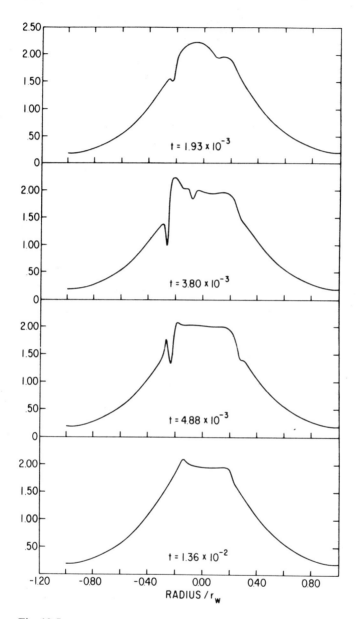

Fig. 10.5
Cross section of the longitudinal current density profile in response to the $m=1$ resistive tearing mode shown in fig. 10.4. From B. V. Waddell, M. N. Rosenbluth, D. A. Monticello, and R. B. White, *Nuclear Fusion*, 16, 528 (1976).

oscillations and a detailed comparison with experiments has been made by Jahns, Soler, Waddell, Callen, and Hicks (1977) and will be discussed in section 11.1.

An example of a computational result for the nonlinear development of an $m=1$ resistive tearing mode in a circular cylinder is illustrated in figs. 10.4 and 10.5. In the first frame of fig. 10.4, the original magnetic axis of the cylinder has moved to the left and a large $m=1$ island has developed on the right, with its center where the original $q=1$ surface used to be. As the magnetic field lines break at the horns of the crescent-shaped island, they pull away and leave a region of lower magnetic pressure. Then the magnetic pressure from the rest of the island continues to push the original center of the plasma to the left until it is completely destroyed. This has the effect of flattening the current profile and the q-value profile across the center of the plasma as shown in fig. 10.5. The additional heat from the center of the plasma is spilled out beyond the $q=1$ surface. This shoulder of hot plasma rapidly diffuses away and the termperature in the center gradually peaks up again due to Ohmic heating as the sawtooth cycle begins over again. We will return to see how this compares with experiment in section 11.1.

Before the current profile has completely flattened, the $m=1$ island width continues to grow exponentially at close to the linear growth rate. This is completely different from the behavior of the $m \geq 2$ tearing modes discussed in the last section. As the current profile flattens, the q-value becomes greater than one everywhere, the mode suddenly stops growing, leaving only the vortex patterns of the velocity field which subsequently damp out because of viscosity. In figs. 10.4 and 10.5, the exponential growth of the instability has just stopped at the time of the second frame.

10.4 RESISTIVE INTERCHANGE MODE

The second most widely studied resistive instability is called the *resistive interchange mode*. This instability is driven by an inverted density gradient in a gravitational field or an inverted pressure gradient in a curved magnetic field. Hence, the resistive interchange mode is analogous to the Rayleigh-Taylor instability in a plane slab geometry or to such ideal MHD interchange modes as the Suydam, Mercier or ballooning modes in cylindrical or toroidal geometry. While kink or tearing modes are driven by the magnetic potential energy at large, interchange modes are driven by local pressure gradients and field curvature.

In the first part of this section, a simple estimate will be made for the growth rate and boundary layer width of a resistive interchange mode in plane slab geometry. In the second part, a dispersion relation that incorporates both the resistive interchange and tearing modes in realistic toroidal geometry will be described. The growth rate estimate, which comes directly from Furth, Killeen, and Rosenbluth (1963), will be presented in much the same spirit as the estimates made in section 10.2

Consider a plane slab with an inverted density gradient in a gravitational field and with a sheared magnetic field, as described in section 10.2. Two force terms will now be considered in the equation of motion

$$\rho^0 \frac{\partial}{\partial t} v_x^1 = -\rho^1 g - J_z^1 B_* \tag{10.4.1}$$

where the notation has been defined in section 10.2 except for the gravitational acceleration $-g$ and the perturbed density ρ^1 determined by

$$\frac{\partial}{\partial t}\rho^1 = -v_x^1 \frac{\partial}{\partial x}\rho^0(x) \tag{10.4.2}$$

for this incompressible plasma motion. Since v_y^1 is much larger than v_x^1, the best estimate of the inertial effects is made by balancing the rate of change of kinetic energy

$$\tfrac{1}{2}\rho^0 \frac{\partial}{\partial t}[(v_x^1)^2 + (v_y^1)^2] \simeq \rho^0 \gamma (v_x^1)^2/(k\varepsilon)^2 \tag{10.4.3}$$

against the work done by the gravitational field

$$\mathbf{F}_g \cdot v_x^1 = -\rho^1 g v_x^1 \tag{10.4.4}$$

and the work done by the $\mathbf{J} \times \mathbf{B}$ force

$$\mathbf{F}_{\mathbf{J} \times \mathbf{B}} \cdot v_x^1 = -J_z^1 B_* v_x^1 \tag{10.4.5}$$

where

$$J_z^1 \simeq v_x^1 B_*/\eta \tag{10.4.6}$$

within the boundary layer. Balancing these three terms and using (10.2.18), yields an estimate for the growth rate of the resistive inter-

Resistive Instabilities

change mode

$$\gamma \simeq \left(\frac{\eta}{\rho}\right)^{1/3} \left(\frac{\rho' g k}{B'_*}\right)^{2/3} \tag{10.4.7}$$

$$\simeq \tau_R^{-1/3} \tau_A^{2/3} (kg)^{2/3} \left(\frac{\rho' q}{\rho q'}\right)^{2/3}$$

and the boundary layer thickness

$$\frac{\varepsilon}{a} \simeq \tau_R^{-1/3} \tau_A^{2/3} \left(\frac{\rho' g}{\rho}\right)^{1/6} (ka)^{-1/3} \left(\frac{q}{q'a}\right)^{2/3} \tag{10.4.8}$$

where the resistive and Alfvén times, τ_R and τ_A, were defined by (10.2.22) and (10.2.23).

Note that the resistive interchange instability grows like $\eta^{1/3}$ rather than the slower $\eta^{2/5}$ scaling of the tearing mode. This is partially offset by the fact that shear decreases the growth rate of the interchange mode while the tearing mode thrives on shear. Finally, it can be seen that the resistive interchange mode depends upon only the local conditions at the mode rational surface, unlike the tearing mode where Δ' links the behavior in the boundary layer to the ideal MHD solution in the rest of the plasma. For this reason, the resistive interchange is considered to be a local instability—localized near its mode rational surface—although Roberts and Taylor (1965) have pointed out that many contiguous resistive interchange modes, each with its own mode rational surface and nearly equal growth rate, can couple together to form elongated convection cells which twist through the plasma following the magnetic shear. This global coupling of interchange modes may be of importance for the ballooning mode in high beta tokamaks described in sections 8.2 and 8.3.

Question 10.4.1
How does the growth rate of the resistive interchange mode compare with that of a Rayleigh-Taylor instability (3.3.7) localized near $\mathbf{k} \cdot \mathbf{B} = 0$ in a highly sheared magnetic field?

A complete boundary layer analysis of the linearized resistive MHD equations, which combines features of both the tearing mode and resistive interchanges, has been worked out in toroidal geometry. This work was begun by Coppi, Greene, and Johnson (1966) and was continued in the papers of Glasser, Greene, and Johnson (1975, 1976). A clear, coherent review of this approach is given in the lecture notes by

Greene (1976). The analysis is considerably more involved than the derivation of the Mercier criterion given in section 7.2.

In the limit of a large aspect ratio, low beta, circular cross section tokamak, the Glasser, Greene, and Johnson dispersion relation simplifies to

$$\Delta' = 2.12 \frac{V_s}{X_0} Q^{5/4}(1 - \pi D_R/4Q^{3/2}). \qquad (10.4.9)$$

Some explanation of the notation is required: Here, Δ' is a generalization of definition (10.2.9), the discontinuity in the radial derivative of the radial magnetic field perturbation. As in the derivation of the Mercier criterion, the perpendicular component of the displacement is nearly uniform along magnetic field lines close to the mode rational surface and has the radial form

$$\xi \simeq A|r - r_s|^{-s} + B|r - r_s|^{s+1} \qquad (10.4.10)$$

where r_s is the minor radius of the mode rational surface and A, B, and s are determined by solving the ideal MHD equations away from the resistive boundary layer. From the separate solutions inside and outside the radius of the mode rational surface, Δ' is defined by

$$\Delta' \equiv \frac{A_{\text{outer}}}{B_{\text{outer}}} - \frac{A_{\text{inner}}}{B_{\text{inner}}}. \qquad (10.4.11)$$

This definition reduces to (10.2.9) for a plane slab.

The large scale factor V_s/X_0 in (10.4.9) is defined as the ratio of a macroscopic scale length to the resistive boundary layer thickness given by

$$\frac{V_s}{X_0} \simeq \left[\frac{1}{\rho} \left(\frac{n B_{\text{tor}}}{\eta R} \frac{q'}{q} \right)^2 \Big/ (1 + 2q^2) \right]^{1/6}. \qquad (10.4.12)$$

The growth rate in (10.4.9) is normalized to the scale of the resistive interchange growth rate

$$Q \equiv \gamma/Q_0 \qquad (10.4.13)$$

$$Q_0 \simeq \left[\frac{\eta}{\rho} \left(\frac{n B_{\text{tor}} q'}{Rq} \right)^2 \Big/ (1 + 2q^2) \right]^{1/3}. \qquad (10.4.14)$$

Finally, D_R represents the driving term for the resistive interchange mode

$$D_R \simeq \frac{2\mu p'}{B_{\text{tor}}^2 r_s} \frac{q^4}{(q')^2} \left[1 - \frac{1}{q^2} \right.$$
$$\left. + \frac{qq'}{r_s^3} \int_0^{r_s} dr \frac{r^3}{q^2} \left(1 - \frac{2\mu R^2 q^2 p'}{r B_{\text{tor}}^2} \right) \right]. \tag{10.4.15}$$

For all of the above, ρ is the mass density and η is the resistivity at the mode rational surface, n is the toroidal mode number and $q' = dq/dr$.

There are three regimes of instability predicted by (10.4.9). If $D_R > 0$, the term in the parentheses on the right in (10.4.9) can be nearly 0 (since V_s/X_0 is very large) and the dispersion relation predicts a resistive interchange mode which grows like $Q \simeq (\pi D_R/4)^{2/3}$ or $\gamma \sim \eta^{1/3}$. Note that the driving term D_R is essentially the same as the driving term for the Mercier instability, but without the shear stabilization term. Shear slows down the resistive interchange mode but it does not stabilize it.

If $D_R = 0$ and $\Delta' > 0$, the dispersion relation (10.4.9) predicts the usual tearing mode with

$$\gamma \sim Q_0(X_0/V_s)^{4/5} \sim \eta^{3/5}(q'/q)^{2/5}. \tag{10.4.16}$$

Finally, if $D_R < 0$ and Δ' is sufficiently positive

$$\Delta' > 1.54(V_s/X_0)|D_R|^{5/6} \sim \eta^{-1/3}, \tag{10.4.17}$$

then all three terms in the dispersion relation are important, and the unstable solutions are complex valued. This overstable solution resembles the behavior of Mirnov oscillations. How a complex frequency comes about even in the absence of Hall or finite Larmor radius effects is not entirely clear.

A more general dispersion relation, applicable to arbitrary geometry, and more details on its evaluation can be found in the paper by Glasser, Greene, and Johnson (1976). It is found that toroidicity has a stabilizing influence on tearing modes, although current profile is still the most important factor in determining tearing mode stability. Finite pressure and compressibility also play an important part in these instabilities.

The resistive interchange mode can drive Bénard convection cells in a plane slab, even with magnetic shear (Dagazian and Paris (1977)). Very little is known about the nonlinear development of the mode under these conditions.

10.5 SUMMARY

Under the influence of an externally applied magnetic field perturbation or any of the instabilities discussed below, a resistive plasma tends to break up into helical filaments called magnetic islands. For a given radial magnetic field perturbation, the island width is given by (10.1.9) in a plane slab or by (10.1.14) in a circular cylinder.

A general dispersion relation for the linear growth of $m \geqslant 2$ resistive instabilities is given by (10.4.9). Using simpler models, it was determined that a large pressure gradient in a curved magnetic field or an inverted density gradient in a gravitational field can drive resistive interchange modes with growth rate scaling like $\eta^{1/3}$. Plasma current can drive resistive tearing modes with growth rate scaling like $\eta^{3/5}$, as in (10.2.21). The nonlinear growth of the tearing mode island width is given approximately by (10.2.32), up to and including saturation.

The special case of an $m=1$ resistive tearing mode has growth rate (10.3.1), or (10.3.3) with finite Larmor radius effects included. This mode saturates only after the island has swallowed up the center of the plasma.

10.6 REFERENCES

The classic paper in this field is

H. P. Furth, J. Killeen, and M. N. Rosenbluth, *Phys. Fluids*, 6, 459–484 (1963); also reprinted in *MHD Stability and Thermonuclear Containment*, A. Jeffrey and T. Taniuti, ed. (New York: Academic Press, 1966).

A recent clear review of the mathematical theory of linear resistive instabilities is given by

J. M. Greene, "Introduction to Resistive Instabilities", report LRP 114/76 (Lausanne, Switzerland, 1976).

For a discussion and illustration of magnetic islands, see

A. I. Morozov and L. S. Solov'ev, *Reviews of Plasma Physics*, 2, 1–101 (1966).

J. M. Finn, *Nuclear Fusion*, 15, 845–854 (1975).

P. Chrisman, J. Clarke, and J. Rome, report ORNL/TM-4501 (Oak Ridge, 1974).

S. Matsuda and M. Yoshikawa, *Jap. J. Appl. Phys.*, 14, 87–94 (1975).

M. Vuillemin and C. Gourdon, report EUR-CEA-FC-393 (Fontenay-aux-Roses, France, 1967).

This last reference has a beautiful illustration (reprinted in Mercier and Luc's book) of magnetic islands and apparently ergodic regions in an analytically prescribed vacuum magnetic field.

The reconnection of magnetic field lines is studied by

J. B. Taylor, *Phys. Rev. Letters*, 33, 1139–1141 (1974).

J. D. Jukes, IAEA Berchtesgaden Conference, 1, 479–488 (1976).

B. B. Kadomtsev, *Sov. J. Plasma Phys.*, 1, 389–391 (1975).

B. B. Kadomtsev, IAEA Berchtesgaden Conference, 1, 555–565 (1976).

A. B. Rechester and T. H. Stix, Phys. Rev. Letters, 36, 587–591 (1976).

T. H. Stix, Phys. Rev. Letters, 36, 521–524 (1976).

M. N. Rosenbluth, R. Z. Sagdeev, J. B. Taylor, and G. M. Zaslavski, Nuclear Fusion, 6, 297–300 (1966).

After Furth, Killeen, and Rosenbluth, tearing modes have been studied by

H. P. Furth, P. H. Rutherford, and H. Selberg, Phys. Fluids, 16, 1054–1063 (1973).

A. H. Glasser, H. P. Furth, and P. H. Rutherford, Phys. Rev. Letters, 38, 234–237 (1977).

R. J. Hastie, A. Sykes, M. Turner, and J. A. Wesson, Nuclear Fusion, 17, 515–523 (1977).

P. H. Rutherford, Phys. Fluids, 16, 1903–1908 (1973).

R. B. White, D. A. Monticello, M. N. Rosenbluth, and B. V. Waddell, Phys. Fluids, 20, 800–805 (1977); IAEA Berchtesgaden Conference, 1, 569–577 (1976).

D. Biskamp and H. Welter, IAEA Berchtesgaden Conference, 1, 579–590 (1976).

J. M. Finn, Phys. Fluids, 20, 1749–1757 (1977).

J. F. Drake and Y. C. Lee, Phys. Fluids, 20, 1341–1353 (1977); Phys. Rev. Letters, 39, 453–456 (1977).

P. H. Rutherford and H. P. Furth, MATT-872 (Princeton, 1971).

J. A. Dibiase and J. Killeen, report UCRL-78796 (Lawrence Livermore Laboratory, 1976); to appear in J. Comp. Phys.

J. A. Wesson, Nuclear Fusion, 6, 130–134 (1966).

E. M. Barston, Phys. Fluids, 12, 2162–2174 (1969).

The $m=1$ resistive instability is treated by

B. Coppi, R. Galvão, R. Pellat, M. Rosenbluth, and P. Rutherford, Sov. J. Plasma Phys., 2, 533–535 (1976).

B. V. Waddell, G. Laval, and M. N. Rosenbluth, report ORNL/TM-5968 (Oak Ridge, 1977).

M. N. Bussac, D. Edery, R. Pellat, and J. L. Soule, IAEA Berchtesgaden Conference, 1, 607–613 (1977).

B. V. Waddell, M. N. Rosenbluth, D. A. Monticello, and R. B. White, Nuclear Fusion, 16, 528–532 (1976).

In addition to Greene's review article and Furth, Killeen and Rosenbluth, cited above, resistive interchange modes are studied by

A. H. Glasser, J. M. Greene, and J. L. Johnson, Phys. Fluids, 19, 567–574 (1976).

A. B. Mikhailovskii, Nuclear Fusion, 15, 95–102 (1975).

J. L. Johnson and J. M. Greene, Plasma Physics, 9, 611–629 (1967).

B. Coppi, J. M. Greene, and J. L. Johnson, *Nuclear Fusion*, *6*, 101–117 (1966).

K. V. Roberts and J. B. Taylor, *Phys. Fluids*, *8*, 315–322 (1965).

R. Y. Dagazian, *Phys. Fluids*, *19*, 169–170 (1976).

R. Y. Dagazian and R. B. Paris, *Phys. Fluids*, *20*, 917–927 (1977).

Other references cited in the text are

M. N. Rosenbluth, R. Y. Dagazian, and P. H. Rutherford, *Phys. Fluids*, *16*, 1894–1902 (1973).

G. L. Jahns, M. Soler, B. V. Waddell, J. D. Callen, and H. R. Hicks, "Internal Disruptions in Tokamaks", to appear in *Nuclear Fusion*.

J. D. Callen and G. L. Jahns, *Phys. Rev. Letters*, *38*, 491–494 (1977).

11
Comparison Between Theory and Experiment

At least five kinds of large-scale instabilities are observed in toroidal confinement devices.
1. Sawtooth oscillations or minidisruptions are observed as relaxation oscillations in the soft X rays emitted from the central part of the plasma column.
2. Mirnov oscillations are regular magnetic field oscillations which can be detected outside the plasma column corresponding to moving helical structures observed in the soft X-ray signal from near the edge of the plasma.
3. The disruptive instability is an abrupt expansion of the temperature and current profiles accompanied by a large negative voltage spike, a sudden loss of runaway electrons, and many other effects.
4. The $m=1$ kink instability, which occurs when $q_{edge} < 1$, throws the plasma against the wall.
5. The $m=0$ sausage instability, which occurs when there is virtually no toroidal magnetic field present, produces large local electric fields and rapidly destroys plasma confinement.

The first two instabilities are benign and are observed during normal tokamak operation. The third instability can terminate the plasma discharge and damage the walls, but it can be avoided by carefully selecting the operating parameters. The last two instabilities were a problem in the early days, but they are easily avoided by using a strong toroidal magnetic field. There are other kinds of instabilities and other ways to organize the experimental observations, but this division is generally used by tokamak experimentalists today.

Sawtooth oscillations, Mirnov oscillations, and the disruptive instability will be the main subjects in this chapter. An attempt will be made to describe a scenario for each phenomenon, based on the best available theoretical model, and indicate where further work is needed. While a complete comparison between theory and experiment is beyond the scope of this chapter, references will be given to many papers where such comparisons are made. In order to avoid unnecessary repetition, the reader is urged to review the description of the experimental observation of these instabilities in chapter 1, as necessary.

11.1 SAWTOOTH OSCILLATIONS

A very convincing scenario for sawtooth oscillations has been developed by Jahns, Soler, Waddell, Callen, and Hicks (1977). Parts of this model were previously suggested by Kadomtsev (1975), Sykes and Wesson (1976), and others. According to their model, sawtooth oscillations result from the competition between two effects. First, the toroidal current density tends to concentrate at the magnetic axis because the center of the plasma generally heats up more rapidly than the edge— this is called a "thermal instability" (Furth et al. (1970)). When the current concentrates so much that the q-value drops below unity at the magnetic axis, an $m=1$, $n=1$ resistive tearing mode becomes unstable and flattens the temperature and current profiles over the center of the plasma. As we shall see, the $m=1$ instability can grow explosively fast, leading to a sudden drop in temperature at the center of the plasma and a correspondingly sudden increase in the temperature just outside the $q=1$ surface. Then, returning to a slower time scale, the excess temperature beyond the $q=1$ surface diffuses away and the center of the plasma heats up again, leading to another sawtooth oscillation.

This section will summarize the work of Jahns, Soler, Waddell, Callen, and Hicks (1977), and Callen and Jahns (1977), which is specifically devoted to finding quantitative agreement between the theoretical prediction and the experimental observation of sawtooth oscillations. First, we will consider those effects that pertain to just the $m=1$ mode, and then we will consider the transport effects that control most of the axisymmetric $m=0$ part of the sawtooth oscillations.

As a prerequisite for the $m=1$, $m=1$ mode to be unstable, a $q=1$ surface must lie within the plasma. For peaked current profiles, the

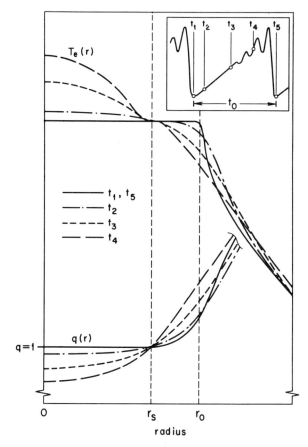

Fig. 11.1
Model for the evolution of electron temperature and q-value profiles during a sawtooth oscillation. From G. L. Jahns, M. Soler, B. V. Waddell, J. D. Callen, and H. R. Hicks, "Internal Disruptions in Tokamaks," to appear in *Nuclear Fusion* (1978).

q-value is smallest at the magnetic axis and increases monotonically with radius to the edge of the plasma. Therefore, as the current increases, or becomes more centrally peaked, the $q=1$ surface will first appear at the magnetic axis and then gradually move out in radius. For any axisymmetric toroidal plasma, the q-value at the magnetic axis is given by

$$q_{\text{axis}} = B_\varphi \bigg/ R \left| \frac{\partial}{\partial R}(RB_y) \frac{\partial}{\partial y}(RB_R) \right|^{1/2}, \tag{11.1.1}$$

in general, or by

$$q_{\text{axis}} \simeq B_\varphi / R \tfrac{1}{2} J_\varphi \tag{11.1.2}$$

for a low beta plasma with circular cross section. It is generally impossible to make a direct experimental measurement of the magnetic field or current density within a hot plasma, but the central current density can be inferred from Ohm's law with Spitzer resistivity (2.6.18), using measured values for the toroidal electric field ($E_\varphi \simeq E_{\varphi 0} R_0/R$ in steady state) and the electron temperature (from Thompson scattering of a laser beam) and using the best estimate for the impurity concentration ($Z_{\text{effective}}$) inferred from other measurements. The estimate of $Z_{\text{effective}}$ is the weakest link in the chain. However, there is remarkable agreement that the condition $q_{\text{axis}}=1$ coincides with the onset of sawtooth oscillations over a wide range of tokamak conditions.

Consider now the time dependence of sawtooth oscillations. The TFR group (1976) has established that the X-ray signal is primarily due to changes in the electron temperature and not due to changes in density or other variables. Jahns et al. (1977) and others have shown that the slowly rising part of the sawtooth oscillations near the center of the plasma can be accounted for by Ohmic heating and similar energy transport processes. The final step, then, is to show how the $m=1$ mode leads to a sudden temperature drop at the end of each sawtooth oscillation.

Since the growth rate of the $m=1$ mode is strongly affected by shear at the $q=1$ surface, Jahns et al. (1977) developed a model for the time evolution of the $T_e(r)$, $J_z(r)$, and $q(r)$ profiles near the center of the plasma. This model is illustrated in fig. 11.1, where the deformation of the $q(r)$ profile has been exaggerated for clarity. It is assumed that the position of the $q=1$ surface remains fixed at $r=r_s$ during the whole sawtooth oscillation. As the electron temperature and the toroidal current derived from Ohm's law peak at the magnetic axis, the shear $d\ln q/d\ln r$ at r_s increases dramatically. For example, if the electron

Comparison Between Theory and Experiment

temperature increases linearly in time with the form

$$T_e(r,t) = T_{e0}\left\{1 + \frac{t}{\tau_h}[1 - 3(r/r_s)^2 + 2(r/r_s)^3]\right\}, \tag{11.1.3}$$

which is consistent with $\partial T_e/\partial r|_{r_s} = 0$ and T_e fixed at r_s, then Faraday's law together with Spitzer resistivity

$$\frac{\partial B_\theta}{\partial t} \simeq \frac{\partial}{\partial r}\eta J_z \simeq -\tfrac{3}{2}\eta_0 J_{z0}\frac{1}{T_e}\frac{\partial T_e}{\partial r}$$

implies

$$\left.\frac{\partial}{\partial t}\frac{\partial \ln q}{\partial \ln r}\right|_{r_s} \simeq \frac{q\eta_0 J_{z0} r}{B_\theta}\left.\frac{t}{\tau_h}\right|_{r_s}. \tag{11.1.4}$$

It follows that the shear at $r = r_s$ increases quadratically in time

$$\frac{\partial q}{\partial r} = \left.\frac{\partial q}{\partial r}\right|_{t=0} + \left.\frac{q\eta_0 J_{z0} q}{2B_\theta \tau_h}\right|_{r_s} t^2 \tag{11.1.5}$$

where the heating rate at the magnetic axis, $1/\tau_h$, is given by

$$\tau_h \simeq \tfrac{3}{2} n_{e0} T_{e0}/\eta_0 J_{z0}^2 \tag{11.1.6}$$

when Ohmic heating dominates.

The growth rate of the $m = 1$ resistive tearing mode with finite Larmor radius effects included scales like $(q'(r_s))^2$, according to (10.3.3) and (10.3.1). The time development of this mode $\exp(\int^t dt \gamma(t))$ during each sawtooth oscillation is predicted to make a transition from simple exponential growth, $\exp(\gamma_0 t)$, for as long as the initial value of the shear dominates in (11.1.5), to an explosively fast growth, $\exp(\gamma_4 t^5)$, after the quadratic term dominates in (11.1.5). This prediction is confirmed by the experimental observations. Furthermore, the time scale for the sudden temperature drop is consistent with this particular mode and is observed to be clearly inconsistent with corresponding estimates made with the ideal MHD $m = 1$ internal kink mode or with the resistive $m = 1$ tearing mode without finite Larmor radius effects. Finally, using the assumption that the $m = 1$ island width is initially equal to the tearing layer width (10.2.19), and that the $\exp(\gamma_4 t^5)$ behavior dominates most of the time evolution, a surprisingly good estimate can be made for the repetition time for each sawtooth oscillation

$$\bar{t} \simeq \left(\frac{r_s}{a}\right)^{6/5}\left(\frac{T_i}{T_e}\right)^{1/5}\left(\frac{\tau_A}{ka}\right)^{2/5}\tau_h^{2/5}\omega_*^{2/5}\tau_R^{3/5} \tag{11.1.7}$$

where $r_s = r_{q=1}$, a is the radius of the plasma, τ_R, τ_A, and τ_h are given by (10.2.22), (10.2.23), and (11.1.6), and it is assumed that $\omega_{*i} = \omega_{*e} \equiv \omega_{*}$ from (10.3.2) and (10.3.5)

The spatial structure of the oscillating $m = 1$ X-ray signal is observed to be strongest just inside the $q = 1$ surface, dropping off rapidly at larger radii. Computer simulations of the $m = 1$, $n = 1$ resistive tearing mode, such as the ones described in section 10.3 and illustrated in figs. 10.4–10.5, reproduce this structure quite well as a function of radius. The rotation frequency, however, has not yet been accurately predicted.

The heat which is removed from the center of the plasma during its sudden temperature drop is deposited in a thin layer just beyond the $q = 1$ surface, where it subsequently diffuses away. This local heat pulse provides an excellent opportunity to measure the local heat conductivity across the flux surfaces beyond the $q = 1$ surface. The propagation of this heat pulse away from the $q = 1$ surface can be seen quite clearly in the bottom five oscilloscope traces in fig. 11.2. After a careful analysis of the experimental data, Callen and Jahns (1977) were able to show that the heat pulse propagates through this region by diffusion alone with a nearly uniform coefficient of electron heat conductivity. Taking into

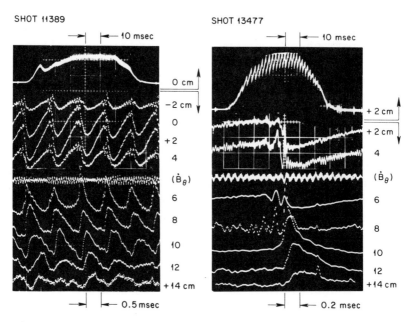

Fig. 11.2
Oscilloscope traces of the soft X-ray signal from sawtooth oscillations at different radii (see arrangement of detectors in fig. 1.6). From J. D. Callen and G. L. Jahns, *Phys. Rev. Letters*, **38**, 491 (1977).

account the fact that the heat pulse propagates inward as well as outward, since the temperature perturbation is negative inside the $q=1$ surface and positive outside, Soler (see Jahns et al., 1977) demonstrated that the heat conductivity derived from soft X-ray observations is close to the value estimated from global energy confinement considerations, within the range of experimental error.

At this point it can be seen that there is at least one consistent model that accounts for most of the experimental observations of sawtooth oscillations that can be used to predict their behavior over a wide range of circumstances. In general, it is believed that sawtooth oscillations by themselves are relatively benign since they mix up only the central part of the plasma, leaving the edge relatively untouched. However, sawtooth oscillations may prove to be dangerous if they play a role in the major disruptive instability (see section 11.3), or if they get too large.

At least two features of sawtooth oscillations are not reliably predicted by the theories at hand. It is not yet entirely clear what determines the position of the $q=1$ surface within the plasma, and why it remains fixed during the sawtooth oscillations. Also, the origin of the rotation frequency of the $m=1, n=1$ helical structure is still the subject of debate; it may be due to diamagnetic effects (mainly from the Hall term in (2.2.16) and gyrovisiosity) or the rotation of the torus as a whole. This will be discussed in the next section on Mirnov oscillations.

Question 11.1.1
For a given q-value at the edge of the plasma, what is the maximum possible radius of the $q=1$ surface?

Question 11.1.2 (Kadomtsev (1975); Smith (1976))
Let $S(r)$ be the source function of the soft X-ray emission. Suppose the sawtooth oscillation changes $S(r)$ from a roughly parabolic form, $S(r) \simeq S_0(1 - r^2/b^2)$, where b is less than the radius of the plasma, to a flat form over $0 < r < c$ such that the volume average of $S(r)$ is fixed and the value at the $q=1$ surface, $S(r_s)$, is fixed. What is the value of c relative to r_s?

11.2 MIRNOV OSCILLATIONS

It is now generally believed that Mirnov oscillations are the result of nonlinearly saturated magnetic islands produced by resistive tearing modes, as described in section 10.2. Strong support for this model comes from the spatial structure observed in the soft X-ray signal emitted from

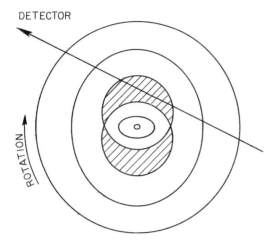

Fig. 11.3
Rotating $m=2$ island structure and the line of sight to an X-ray detector for question 11.2.1.

the region between the $q=1$ surface and the edge of the plasma. The analysis of the soft X-ray signal by von Goeler (1975), for example, is consistent with a flattening of the temperature profile within helical magnetic islands having a nearly steady state width of less than a few tenths of the plasma radius. A cross section of this temperature profile is illustrated in fig. 11.3. The nonlinear saturation width of the magnetic islands produced by tearing modes is predicted to be in this range by White, Monticello, Rosenbluth, and Waddell (1977), as described by eqs. (10.2.32)–(10.2.34) in section 10.2. The perturbed current in these islands flows parallel to the island helicity in the positive direction at the 0-point and in the negative direction at the x-point, as illustrated in fig. 10.3. It is the magnetic field perturbation produced by these helical currents that is detected as Mirnov oscillations at the edge of the plasma.

Question 11.2.1 (von Goeler (1975))
Suppose the X-ray source function is peaked at the center of the plasma and flattened within each magnetic island, as shown in fig. 11.3. What would be the signal from X-ray detectors looking through different chords of the plasma if this whole structure were rotating around the center of the plasma?

The stability and growth rate of tearing modes depends rather sensitively on the current profile of the plasma (which determines the parameter Δ' (10.2.9) used throughout sections 10.2 and 10.4). Rutherford (1973), for example, predicts that a succession of ..., $m=4$, $m=3$, $m=2$ tearing modes with $n=1$ should be observed as the current profile shrinks. This agrees with the succession of Mirnov oscillations observed during the initial phase of a tokamak discharge, as illustrated in fig. 1.5.

There are several models which predict the rotation rate of the helical structures producing Mirnov oscillations. One model, based on the Hall effect, was first suggested by Ware (1961). Roughly speaking, the Hall effect appears in Ohm's law (2.6.16) in the form

$$E \sim -\mathbf{v} \times \mathbf{B} + \mathbf{J} \times \mathbf{B}/ne \qquad (11.2.1)$$

$$\sim -\mathbf{v}_i \times \mathbf{B} + (\mathbf{v}_i - \mathbf{v}_e) \times \mathbf{B}$$

$$\sim -\mathbf{v}_e \times \mathbf{B}$$

so that Faraday's law

$$\frac{\partial \mathbf{B}}{\partial t} = -\nabla \times \mathbf{E} \sim \nabla \times (\mathbf{v}_e \times \mathbf{B}) \qquad (11.2.2)$$

implies that magnetic field lines move with the electron velocity $\mathbf{v}_e$ rather than with the fluid or ion velocity $\mathbf{v} \simeq \mathbf{v}_i$. The actual prediction is somewhat more complicated since the full Hall effect involves electron and ion pressures separately. However, this simple argument indicates that the motion of field lines should show up as a toroidal rotation of the magnetic islands in the direction of the electron current flow. This is roughly what is observed in experiments. The inference of toroidal rather than poloidal rotation follows from the observation that the modes are often phase locked together under those conditions where combinations of $m=2$, $n=1$ and $m=3$, $n=1$ helices are present simultaneously. More sophisticated analyses by Rutherford and Furth (1971) and by Glasser, Greene, and Johnson (1976) indicate that the island structure should rotate at roughly the electron diamagnetic frequency, which is also consistent with experimental observations. A final possibility is that the island rotation is altered by the toroidal rotation of the plasma as a whole, as predicted and experimentally observed by Sigmar, Clarke, Neidigh, and Vander Sluis (1974). The inconclusive comparison between the observed rotation and theoretical predictions is somewhat complicated by the fact that the rotation of Mirnov oscillations can be stopped altogether by applying a resonant helical magnetic field produced by external coils, described in the next section.

It is clear that accurate profile determinations are needed from the experiments, and predictions with realistic profiles are needed from the theories before a more conclusive comparison between theory and experiment can be made.

11.3 THE DISRUPTIVE INSTABILITY

A remarkable experiment on disruptive instabilities was carried out on the Pulsator tokamak in Garching, Germany. This tokamak is equipped with a set of $m=2$, $n=1$ helical coils around the plasma. As the current in these helical coils is increased, it is observed first that the rotation of Mirnov oscillations stops and that the occurrence of the disruptive instability, in an otherwise disruptive plasma, is delayed a significant amount. From similar experiments on the ATC tokamak in Princeton, it is inferred that the rotation of the Mirnov oscillations is altered and not the amplitude, since the oscillations resume at their normally expected amplitude when the helical field is turned off. Returning to the Pulsator experiment, as the external helical current is increased further, a sharp threshold is observed at which a disruptive instability is triggered, even in a discharge that is not otherwise disruptive. Hence, an applied helical magnetic field can either delay or

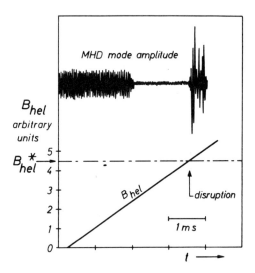

Fig. 11.4
An applied helical magnetic field first stops the rotation of Mirnov oscillations and then induces a disruptive instability in the Pulsator tokamak, Garching. From F. Karger et al., IAEA Tokyo Conference, *1*, 207 (1974).

trigger a disruptive instability, depending upon the amplitude of the applied field.

An explanation of this behavior was suggested by Lackner and Karger (1977). Their work is based on the idea that a disruptive instability is produced when an $m=2$, $n=1$ magnetic island gets too wide within the plasma. The $m=2$ island interacts with the limiter or the $m=1$ island associated with sawtooth oscillations or other induced $m > 2, n \geqslant 1$ islands, in a process that will be described in more detail later. The externally applied helical field makes an $m=2$ island with controllable width, in contrast to the natural island produced by a tearing mode. If the controlled island is made too wide, it interacts with the limiter, cooling the edge of the plasma and forcing all the current to run in such a narrow channel that the $m=1$ mode becomes unstable at a large radius, thus inducing a disruption. If the controlled island is made smaller, it flattens the current profile at the $q=2$ surface and reduces the shear there so that the $m=2$, $n=1$ tearing mode grows more slowly, thus delaying the disruption. A natural disruption occurs only when plasma conditions allow the $m=2$ tearing mode island to grow too wide.

The idea that the disruptive instability is caused by an $m=2$ magnetic island interacting with other structure within the plasma has been suggested by many authors and has gathered increasing support from experimental evidence (Pulsator group (1976); T.F.R. group (1977)). For example, detailed analysis of the soft X-ray signal by von Goeler (1975) indicates that the disruptive instability starts with a temperature flattening at the $q=2$ surface and spreads outward in both directions, followed by a negative voltage spike after 200 μsec. Multiple magnetic probe measurements made in the smaller and cooler LT-3 tokamak by Hutchinson (1976) show that the expansion of the current profile begins between the $q=2$ and $q=1$ surfaces, as illustrated in fig. 11.5. Just before disruption, the magnetic signal at the edge of the plasma develops a strong $m=2$ perturbation which rapidly breaks into a more complicated pattern, as shown in fig. 11.6 by Vlasenkov et al. (1974).

Magnetic islands can interact with each other in a number of possible ways. The strongest kind of interaction occurs when islands overlap. Computations by Finn (1975) show that as magnetic islands are brought closer together, an annular region filled with ergodically (or randomly) wandering field lines forms between the interacting islands and abruptly widens to the full region of island overlap. The time dependence of this process is not yet entirely clear. Even while the islands are farther apart, secondary islands form within the original

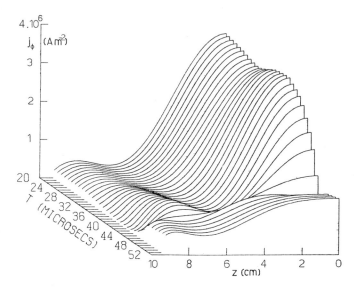

Fig. 11.5
Time evolution of the toroidal current density as a function of minor radius (z) during a disruptive instability in the LT-3 tokamak in Canberra, Australia. From I. H. Hutchinson, *Phys. Rev. Letters*, **37**, 338 (1976).

islands wherever the Fourier components of the mutual perturbation resonate with the *local* q-value within each primary island. Rechester and Stix (1976) show that these secondary islands pile up in a boundary layer around the separatrices of the primary island and the boundary layer rapidly grows wider as the strength of the interaction is increased. In addition to these interactions current-carrying islands can attract each other most strongly at points around the outer edge of the torus where they are closest (Pulsator group (1976); Kadomtsev (1976)). The islands are most likely to join together at these points by a process of field line reconnection allowed by a small plasma resistance.

The first effect of field line reconnection should be a loss of confinement for fast runaway particles. This is in fact observed as a burst of hard X rays in the first of any series of disruptions. The second effect of field line reconnection should be a rapid expansion of the electron temperature profile, since the electron heat conductivity parallel to magnetic field lines is extremely large (2.6.19). This effect shows up in the soft X-ray observations. Finally, the current profile spreads out. A recent theory developed by Carreras, Waddell, and Hicks (1977) indicates that a sufficiently large $m=2/n=1$ tearing mode will nonlinearly drive other tearing modes with odd harmonics (such as

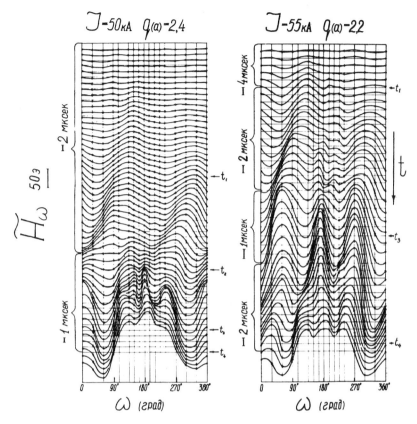

Fig. 11.6
Time dependence of the poloidal magnetic field perturbation as a function of poloidal angle ω, measured from the outer edge of the torus, leading up to a disruptive instability in the T-6 tokamak at the Kurchatov Institute in Moscow. Time proceeds from top to bottom in increments of 1 or 2 μsec, as noted. From V. S. Vlasenkov, V. M. Leonov, V. G. Merezhkin, and V. S. Mukhovatov, IAEA Tokyo Conference, *1*, 33 (1974).

$m=3/n=2$ and $m=5/n=3$) into explosively fast growth. The sizeable perturbed current spikes associated with all these tearing mode islands produce the combined effect of a spread in the total current profile. A mechanism like this is needed to explain the observed rapid expansion of the current profile—much too rapid to be explained by resistive magnetic field diffusion following the expansion of the electron temperature profile. If the plasma recovers from the disruption, these current spikes presumably spread out and merge due to resistivity to form a new axisymmetric equilibrium.

The effect of the spreading profiles is to reduce the poloidal beta and internal inductance so that the toroidal plasma is pushed inward by the applied vertical field (4.7.12). The combined effect of an increase in the minor radius and a decrease in the major radius of the current channel is to produce a mild increase in the toroidal current and a large negative voltage spike (see question 1.2.3). The decrease in the poloidal magnetic field near the center of the plasma produces a large positive electric field there. Hutchinson (1976) speculates that this electric field accelerates a small number of ions to very high energy and these ions are subsequently lost from the plasma.

It is clear then that the basic processes have probably been identified but that the full time dependence of disruptive instabilities has yet to be worked out in detail. It has been difficult to construct a complete picture because the disruptive instability occurs so suddenly and unpredictably after some internal rearrangement. It is remarkable that such detailed experimental measurements have been made. Work is under way to study the disruptions that occur under different conditions described in section 1.2.

11.4 REFERENCES

Sawtooth oscillations are considered by

G. L. Jahns, M. Soler, B. V. Waddell, J. D. Callen, and H. R. Hicks, "Internal Disruptions in Tokamaks," *Nuclear Fusion, 18*, 609–628 (1978).

J. D. Callen and G. L. Jahns, *Phys. Rev. Letters, 38*, 491–494 (1977).

B. B. Kadomtsev, *Sov. J. Plasma Phys., 1*, 389–391 (1975).

B. V. Waddell, M. N. Rosenbluth, D. A. Monticello, and R. B. White, *Nuclear Fusion, 16*, 528–532 (1976).

A. Sykes and J. A. Wesson, *Phys. Rev. Letters, 37*, 140–143 (1976).

H. P. Furth, M. N. Rosenbluth, P. H. Rutherford, and W. Stodiek, *Phys. Fluids, 13*, 3020–3030 (1970).

TFR Group (D. Launois), Seventh European Conf. on Controlled Fusion and Plasma Physics (Lausanne), 2, 1–13 (1975).

R. R. Smith, *Nuclear Fusion*, 16, 225–229 (1976).

TFR Group, report CEA FC 832 (Fontenay-aux-Roses, France, 1976), to appear in *Plasma Physics*; IAEA Berchtesgaden Conf., 1, 279–287 (1976).

W VII A Stellarator team, IAEA Berchtesgaden Conf., 2, 81–93 (1976).

For the theory of Mirnov oscillations, see the references on tearing modes in chapter 10 as well as

S. Von Goeler, Seventh European Conf. on Controlled Fusion and Plasma Phys. (Lausanne), 2, 71–80 (1975).

R. B. White, D. A. Monticello, M. N. Rosenbluth, and B. V. Waddell, *Phys. Fluids*, 20, 800–805 (1977).

P. H. Rutherford, *Phys. Fluids*, 16, 1903–1908 (1973).

A. A. Ware, IAEA Salzburg Conference 1961, *Nuclear Fusion Supplement*, part 3, 869–876 (1962).

P. H. Rutherford and H. P. Furth, report MATT-872 (Princeton, 1971).

A. H. Glasser, J. M. Greene, and J. L. Johnson, *Phys. Fluids*, 19, 567–574 (1976).

D. J. Sigmar, J. A. Clarke, R. V. Neidigh, and K. L. Vander Sluis, *Phys. Rev. Letters*, 33, 1376–1379 (1974).

K. Bol et al., IAEA Tokyo Conference, 1, 83–97 (1974).

The disruptive instability is investigated by

F. Karger et al. (Pulsator group), IAEA Tokyo Conference, 1, 207–213 (1974); IAEA Berchtesgaden Conference, 1, 267–277 (1976).

K. Lackner and F. Karger, "Stabilization of the Disruptive Instability by Resonant Helical Windings," (Garching, 1976).

Equipe TFR, *Nuclear Fusion*, 17, 1283–1296 (1977).

I. H. Hutchinson, *Phys. Rev. Letters*, 37, 338–341 (1976).

D. B. Albert and A. H. Morton, *Nuclear Fusion*, 17, 863–868 (1977).

V. S. Vlasenkov, V. M. Leonov, V. G. Merezhkin, and V. S. Mukhovatov, IAEA Tokyo Conference, *Nuclear Fusion Supplement*, 1–5 (1974).

J. M. Finn, *Nuclear Fusion*, 15, 845–854 (1975).

A. B. Rechester and T. H. Stix, *Phys. Rev. Letters*, 36, 587–591 (1976).

B. B. Kadomtsev, IAEA Berchtesgaden Conference, 1, 555–565 (1976).

B. Carreras, B. V. Waddell, and H. R. Hicks, report ORNL/TM-6096, (Oak Ridge, 1977).

Space did not permit a comparison between the theory and experimental observation of the $m=1$ kink instability. Good agreement has been found by

H. A. B. Bodin, A. A. Newton, G. H. Wolf, and J. A. Wesson, *Phys. Fluids*, 13, 2735–2746 (1970).

K. S. Thomas, *Phys. Fluids*, *15*, 1658–1666 (1972).

J. P. Friedberg, *Phys. Fluids*, *13*, 1812–1818 (1970).

E. Fünfer, M. Kaufmann, W. Lotz, and J. Neuhauser, IAEA Madison Conference, *3*, 189–200 (1971).

M. D. Kruskal, J. L. Johnson, M. B. Gottlieb, and L. M. Goldman, *Phys. Fluids*, *1*, 421–429 (1958).

Also, many papers in the Geneva Conference (1958) were concerned with the stabilization of the $m=1$ kink mode.

In shock heated experiments, the predicted finite Larmor radius stabilization of the $m=2$ mode was demonstrated by

J. Neuhauser, M. Kaufmann, H. Röhr, and G. Schramm, *Nuclear Fusion*, *17*, 3–12 (1977).

Finally, the effect of the $m=0$ sausage instability was investigated by

O. A. Anderson, et al., *Phys. Rev.*, *110*, 1375–1387 (1958).

Appendix A: Comments on the Questions

The comments provided below are intended to help the reader find answers to the questions that appear throughout the text, identified by section number. In most cases the answers involve a simple insight and some rudimentary derivation. However, in some cases, the more challenging questions are intended to extend the material in the text and to provoke interest for future work in the field. Not everyone will agree with my answers.

1 INTRODUCTION

1.1.1

The pressure is $nT = .16$ Joule/cm^3 = 1.6 atmospheres corresponding to a magnetic field of .63 tesla. Only the pressure at the ends of the column has to be absorbed by the structure. For a toroidal plasma, only the hoop force discussed in section 4.7 must be absorbed by the structure under the conditions described here.

1.1.2

$B_{tor} = 9\, B_{pol} = 5.7$ tesla corresponding to roughly 130 atmospheres of pressure against the coils. It is feasible, but very difficult, to make coils with a meter radius to produce such fields.

1.2.1

The sum of the electromotive forces equals the sum of the "IR drops" due to current and resistance around each closed circuit. For the closed loop of resistive wire, $-d\Phi/dt = IR$ determines the current I, where R is the integrated resistance around the wire. For the closed loop consisting of the voltage V across the voltmeter and the segment R_1 closest to the meter, $V = IR_1$ determines the measured voltage. The same voltage is determined from the loop consisting of the voltmeter leads and the segment R_2 away from the meter $-d\Phi/dt = IR_2 + V$, where $R_1 + R_2 = R$.

1.2.2

There is an electromotive force across the gaps in the conducting shell, and there is a resistive voltage drop continuously distributed around the plasma. If the resistance is axisymmetric, the plasma can remain axisymmetric.

1.2.3

If there were an essentially uniform current distribution within radius a (the effective plasma radius), the magnetic energy for a cylindrical plasma of length $2\pi R$ would be

$$(1/2\mu)\int d^3x B^2 = (\mu I^2/2)[\tfrac{1}{4} + \ln(R/a)].$$

Suppose $a \to a + \delta$ and $R \to R - \Delta$ then the change in magnetic energy is

$$\mu I^2/2 \left\{ -(\tfrac{1}{4} + \ln(R/a))\Delta - \Delta - \delta\frac{R}{a} \cdots \right\}.$$

Hence an increase in a and a decrease in R decreases the magnetic energy due to plasma current; the plasma does work on the transformer.

1.2.4

Up to a scale factor, the field outside of the plasma is independent of r_0 in the example given, and there is no way to determine the radial distribution of currents in a multipole just from external magnetic field measurements. This is most easily demonstrated for a straight multipole configuration ($k=0$ in the example given).

2 THE MHD EQUATIONS

2.2.1

The field lines that close upon themselves are in one-to-one correspondence with the rational numbers while the field lines that do not close upon themselves (that cover a surface ergodically) correspond to the irrational numbers. Hence, virtually 100% of the field lines do not close upon themselves.

2.2.2

This problem is analogous to giving the vorticity ($\omega = \nabla \times \mathbf{v}$) over a region and then asking for the velocity. A uniform vorticity corresponds to a rigid rotor. But where is the center of the rotor? The boundary conditions are essential, however remote.

2.2.3

The velocity in eq. (2.2.8) refers to the one frame of reference in which the resistivity is zero. It does not refer to the electron or ion velocities.

2.3.1

If a vacuum surrounds the torus, the new magnetic field will be a dipole field around the torus well separated from the local field between the pole pieces. Field lines can break and reconnect in a vacuum because arbitrary closed contours in a vacuum can support electromotive forces.

On the other hand, if a perfectly conducting fluid surrounds the torus, field lines cannot break and reconnect, and the magnetic field must stretch from one pole piece out to the torus and back again. The torus continues to resist being pulled away, and the energy in the field continues to grow as the torus is pulled.

If there is no current in the torus initially, currents will form as the torus is being pulled away before the flux has a chance to change. In the limit of zero resistivity, just the hint of an electromotive force is enough to induce a current.

2.3.2

Since the fluid flow is not continuous, in the sense of adjacent fluid elements remaining adjacent, the magnetic field may change topology.

2.3.3

As the magnetic field is pushed aside, the energy stored in the field increases at the expense of the kinetic energy in the pendulum. If there is enough kinetic energy to force the magnetic field around the pendulum, the pendulum will continue past and regain its kinetic energy until the magnetic field is restored to its original form.

2.3.4

An electric field forms in the ionosphere as the ionized gas flows through the earth's magnetic field.

2.4.1

The inward magnetic pressure is

$$-\frac{1}{2\mu}\nabla_\perp B^2 = -\frac{\hat{\mathbf{r}}}{2\mu}\frac{\partial}{\partial r}B^2(r)$$

and the tension due to magnetic curvature is

$$\frac{1}{\mu}B^2\kappa = -\frac{\hat{\mathbf{r}}}{\mu r}B_\theta^2(r).$$

When J_z and B_z are both uniform, the contributions are equal. Changing the B_z-profile only changes the magnetic pressure contribution—diamagnetism increasing it and paramagnetism decreasing it. For a centrally peaked J_z-profile, the contributions are equal at $r = 0$ and the magnetic pressure decreases more rapidly than the magnetic tension away from $r = 0$.

2.4.2

Every magnetic field must be divergence-free, $\nabla \cdot \mathbf{B} = 0$. The one described in this question is not.

2.5.1

See page 200 of J. D. Jackson (*Classical Electrodynamics*, first edition, New York: Wiley, 1962) for an expression of the conservation of angular momentum for electrodynamic systems in general. The local torque on the plasma is

$$\nabla \times \rho\frac{d\mathbf{v}}{dt} = \nabla \times (\mathbf{J} \times \mathbf{B}) = \mathbf{B}\cdot\nabla\mathbf{J} - \mathbf{J}\cdot\nabla\mathbf{B}.$$

2.5.2

The Poynting flux $\mathbf{E} \times \mathbf{B}$ around the disk is related to the angular momentum of the electromagnetic field. As the fields collapse, this is turned into the angular momentum of the disk.

2.6.1

When the displacement current is included in Ampere's law, $\mathbf{J} = \frac{1}{\mu} \nabla \times \mathbf{B} - \varepsilon \frac{\partial \mathbf{E}}{\partial t}$, $\nabla \cdot \mathbf{J}$ is no longer zero and the equations are consistent. Without the displacement current, we do not have charge conservation.

3 THE RAYLEIGH-TAYLOR INSTABILITY

3.1.1

Locally, let $v_y \sim e^{ily}$, or, more precisely, use the WKB approximation. Then, for

$$l \gg \left|\frac{1}{\rho}\frac{\partial \rho}{\partial y}\right|, \quad l \sim \frac{k}{\gamma}\left(\frac{g\rho' - \rho\gamma^2}{\rho}\right)^{1/2}$$

There is a point of accumulation at $\gamma = 0$. When γ is small, the solutions can be localized (in the sense of oscillating rapidly in y) at any level where $\rho'(y) > 0$.

3.1.2

$$v_y(y) = \sin\left(\frac{\pi n y}{L}\right)\exp(-y/2\lambda)$$

$$\gamma^2 = \frac{g}{\lambda}\frac{k^2}{k^2 + \left(\frac{1}{2\lambda}\right)^2 + \left(\frac{\pi n}{L}\right)^2}$$

$n = 1$, $k^2 \to \infty$ most unstable.

3.1.3

$$v_y(y) = \begin{cases} v_y(0+)\sinh k(a-y)/\sinh(ka), & y > 0 \\ v_y(0-)\sinh k(b+y)/\sinh(kb), & y < 0 \end{cases}$$

$$\gamma^2 = \frac{kg(\rho_+ - \rho_-)}{\rho_+ \coth(ka) + \rho_- \coth(kb)}.$$

This problem is not mathematically well posed in the sense that a discontinuous perturbation ($k \to \infty$) grows infinitely fast. This mathematical difficulty may be removed by including viscosity, surface tension, or other nonideal effects in the model. Then the wavelength of the fastest growing instability will be determined by the added effect.

3.1.4

The partial differential equation (3.1.8) is the same as the Euler equation for the variational form (3.1.10). As such, (3.1.8) applies only to the extrema of the variational form. For a given set of boundary conditions, each solution of the Euler equation is an eigenfunction corresponding to one and only one eigenvalue γ. However, permissible test functions in the variational form may be any mix of eigenfunctions so that γ^2 in (3.1.10) may be any sum of eigenvalues with arbitrary coefficients.

3.1.5

Yes, but there will be extra boundary terms in (3.1.10).

3.1.6

Using the exponential density model of question 3.1.2, there is a self similar solution $\xi = \xi(t=0)e^{\gamma t}$, $\gamma = \sqrt{g/\lambda}$ for the displacement $\xi(\mathbf{x},t) \equiv \int_0^t dt\, v(\mathbf{x},t)$ when the acceleration g is constant over a time interval. Clearly, if g is fixed and the time interval is decreased, the final displacement will be decreased. Even if the acceleration g is increased to hold the final velocity (kinetic energy) fixed while the time interval is decreased $v_{\max} = g t_{\max}$, the final displacement will be decreased. Hence, to limit the amplitude, compress as fast as possible.

3.3.1

Yes. This can be demonstrated with the particular example

$$\frac{\partial \mathbf{B}}{\partial t} = -\mathbf{v}\cdot\nabla\mathbf{B} + \mathbf{B}\cdot\nabla\mathbf{v} - \mathbf{B}\nabla\cdot\mathbf{v}$$

$$\mathbf{v} = v_y \sin(kx)\hat{\mathbf{y}}$$

$$\mathbf{B}(t=0) = B_{x0}\hat{\mathbf{x}}.$$

3.3.2

$$\gamma^2 = \frac{k(\rho_+ - \rho_-)g - [(k \cdot B_+)^2 \coth(ka) + (k \cdot B_-)^2 \coth(kb)]}{\rho_+ \coth(ka) + \rho_- \coth(kb)}.$$

3.3.3

For a given length of pendulum, the conditions for both an inverted and a normal pendulum to be dynamically stable are

$$\frac{\sqrt{2gl}}{\omega} < a \ll l, \qquad \omega > 2\sqrt{g/l},$$

where a is the amplitude and ω is the driving frequency. These conditions can also be met for a range of lengths

$$\frac{\sqrt{2gl_{max}}}{\omega} < a \ll l_{min}, \qquad \omega > 2\sqrt{g/l_{min}}.$$

A more complete theory might remove the $a \ll l$ condition.

4 MHD EQUILIBRIUM

4.1.1

Except for configurations with a high degree of symmetry, only a small fraction of all possible magnetic fields can satisfy $\mathbf{J} \times \mathbf{B} = \nabla p$. For a simple example, try a uniform B_x-field with a circular region of J_z current in it. Even if you satisfy force balance in the current, it will not be satisfied at the edge of the current.

4.2.1

Open ended devices rely on $p_\parallel \neq p_\perp$. The pressure is not a scalar.

4.3.1

No. For the field line on the inner edge of the torus $q = \lim_{n\to\infty} \frac{n}{0} = \infty$, when n is the number of transits the long way around the torus. For the other field lines that wrap around the outer edge of the same surface,

$$q = \lim_{n\to\infty} \frac{n}{1} = \infty.$$

4.3.2

For all the applications needed here, the q-value on the magnetic axis is the continuous limit of the q-values in the immediate neighborhood of the axis.

4.4.1

Because of the dependence on major radius, no nontrivial solution exists for $R^2 p'(\psi) + II'(\psi) = 0$ in a closed toroidal domain. It will be seen in section 4.7 that at least a vertical field is needed for toroidal equilibrium. A more general proof of the need for externally applied fields is supplied by Shafranov (*Reviews of Plasma Physics*, 2, 107 (1966)) using the virial theorem.

4.4.2

No. For a simple proof see *Nuclear Fusion*, 13, 593 (1973), Appendix A.

4.4.3

The magnetic axis at $R = R_0$ is shifted outward relative to the midpoint between the edges of the plasma at $R_\pm = R_0\sqrt{1 \pm X}$, $y = 0$ where $X = [(\psi_0 - \psi_{\text{edge}})/\alpha R_0^2 \psi_1]^{1/2}$. The magnitude of the poloidal field is the same at both edges. Near the magnetic axis, there can be any elongation. For an analysis of the possible elongations away from the magnetic axis, the separatrices must be inspected. This is done for a number of different cases by Solov'ev (*Reviews of Plasma Physics*, 6, 257–260 (1976)).

4.5.1

There is some question as to whether the external wires can always be rearranged to form a circle around the plasma. If not, there is a difference between pushing and pulling. For example, when walls of current are used to push against the sides of the plasma, as in fig. 4.6b, these walls must be very close to the plasma. Singularities in the magnetic field, corresponding to image currents, must be very close to sustain the elongation of the plasma.

4.5.2

If the quadrupole current is too large, there is no equilibrium—the plasma is simply pulled apart even in the absence of an initial perturbation. The highly elongated states are unstable to collapse or further elongation given a small perturbation.

4.6.1

In the limit of an arbitrarily thin current layer, the magnetic and thermodynamic pressure gradients become infinite while the tension due to magnetic curvature remains finite.

4.6.2

The equilibrium is determined by the equation

$$B_{pol}^2(\theta) = \frac{B_{\varphi in}^2 - B_{\varphi out}^2}{(1 + \varepsilon \cos \theta)^2} + 2\mu p, \quad \text{where } \varepsilon = a/R.$$

The minimum possible poloidal field is reached when $B_{pol}(\theta = 180°) = 0$. Under these conditions $\beta_{pol} \equiv 2\mu p/B_{pol}^2(\theta = \pi/2) = 1/(2\varepsilon - \varepsilon^2)$. Then for a given value of $q_0 = \varepsilon B_\varphi/B_{pol}|_{\theta = \pi/2}$ the maximum value is achieved for

$$\beta \equiv \frac{2\mu p}{B_{\varphi out}^2} \simeq \frac{\varepsilon}{2q^2}.$$

4.6.3

The poloidal flux between the plasma and the wall is $\psi = B_{pol} \, a \ln w/a$ per unit length. Holding B_{pol} fixed, this flux is zero when $a = 0$ and when $a = w$, where a is the radius of the plasma and w is the radius of the wall. The flux is maximum at $a/w = 1/e$. The flux between the elongated plasma and the wall $\psi = B_{pol}(w - a)$ achieves the same maximum value for $a_{min}/w = 1 - 1/e$.

4.7.1

Whether the displacement works on the external circuit or the plasma alone, the radial force needed for equilibrium is the same. The force due to the vertical field is considered in a separate step.

4.7.2

Using (4.7.12) together with $B_{pol} \simeq \mu I_{tor}/2\pi a$ and $l_i \sim 1$ we find $\beta_J \sim 2/\varepsilon$, $\varepsilon \equiv a/R$, when $B_y = B_{pol}$ at the inner edge of the torus. This leads to the limit $\langle \beta \rangle \simeq \beta_J B_{pol}^2/B_{tor}^2 \leqslant 2\varepsilon/q^2$. If the pressure and vertical magnetic field are raised further, the plasma shrinks so that $\langle \beta \rangle$ is reduced.

4.7.3

The pressure can be raised indefinitely if there is no cancellation between B_y and B_{pol} and therefore no separatrix. Essentially, the vertical field must curve outward so that it is weaker at the inner edge of the torus. More complicated fringing fields are then needed for vertical stability.

4.7.4

This is an open question. If you consider the drifts of individual collisionless particles in a toroidal plasma with rotational transform, having a vertical field over only a single quadrant seems reasonable. The experimentalists observe that the quadrant field must be four times stronger than the usual axisymmetric vertical field.

4.7.5

Conservation of flux and entropy makes the outward radial force exerted by the plasma larger for inward displacements and smaller for outward displacements than it would be without conservation. This should enhance horizontal stability.

4.7.6

Conservation of vertical flux through the hole in the torus prevents a complete flip. The vertical stability property stabilizes a small rigid flip displacement and $q > 1$ stabilizes the $m = 1$ helical displacement which looks like the beginning of a flip motion.

5 LINEARIZED EQUATIONS AND THE ENERGY PRINCIPLE

5.1

a. Yes, this perturbation is considered marginally stable in a mathematical analysis, even though it is generally harmless.
b. This is an instability since, with each oscillation, the plasma moves farther away from the equilibrium state.

c. Both the old and the new equilibria are stable since the perturbation does not continue to move the state away from either one.
d. The convection cells represent a dynamic equilibrium which is stable.
e. The turbulent state is not in equilibrium most of the time, and any momentary equilibrium is generally not stable.

5.2.1

Out of the three components of the magnetic field, only two are independent because of the constraint $\mathbf{V} \cdot \mathbf{B} = 0$. Hence, the original MHD equations are equivalent to a sixth order system.

5.3.1

Starting with (2.5.4), write a complete second order perturbation. Now write a second order perturbation of the MHD equations (2.1.3)–(2.1.5) in order to evaluate the time derivatives of the second order pressure $p^{(2)}$ and magnetic field $\mathbf{B}^{(2)}$. Also evaluate $\partial \mathbf{B}^1/\partial t$, use vector identities and the equilibrium condition together with self-adjointness (5.3.2) to derive (5.3.3).

5.3.2

For every eigenvalue γ there is another eigenvalue $-\gamma$, corresponding to an instability for time going backward, with eigenfunction $(\mathbf{v}, \mathbf{B}, p)_{-\gamma} = (\mathbf{v}, -\mathbf{B}, -p)_\gamma$. Each instability is really the combination of these two eigenfunctions

$$(\mathbf{v}^1, \mathbf{B}^1, p^1)(x, t) = (\mathbf{v}, \mathbf{B}, p)_\gamma(x)e^{\gamma t} + (\mathbf{v}, \mathbf{B}, p)_{-\gamma}(x)e^{-\gamma t}$$

so that $p^1(x, t=0) = 0$ and $\mathbf{B}^1(x, t=0) = 0$.

5.3.3

Yes, this perturbation represents a state satisfying equilibrium force balance and it shows up as a marginally stable perturbation in a spectral analysis of the linearized MHD equations.

5.5.1

The perturbations p^1 and $\mathbf{B}^1$ increase linearly with time while the velocity field remains constant in time.

6 CIRCULAR CYLINDER INSTABILITIES

6.1.1

$$q(r) = \frac{2(\nu + 1) r^2}{\mu J_0} \frac{k B_z}{a^2} \frac{1}{1 - (1 - r^2/a^2)^{\nu+1}}$$

$q(a)/q(0) = \nu + 1$.

6.1.2

It may be unlikely for the current profile to change and still give the same field and flux at the edge, but it is certainly possible. For example, start with two given concentric surface currents and then find the magnitude and position of a single surface current that gives the same field and flux at some fixed wall position.

6.2.1

The $m=0$ mode is stable if

$$-\frac{d \ln p}{d \ln r} < \frac{\Gamma}{1 + \tfrac{1}{2}\Gamma\beta} \, ,$$

where $\Gamma = 5/3$ and $\beta \equiv p/\tfrac{1}{2}\mu B^2$ as a function of radius.

6.2.2

$\gamma^2 \simeq I_z B_z k / \langle \rho \rangle$.

6.2.3

From Faraday's law, the perturbed magnetic field b_x changes sign when $\tfrac{1}{2}J_z = k B_z$. The $J_z^0 B_x^1$ term in the equation of motion drives this instability.

6.2.4

There are many terms in the expression for $\mathbf{J}^1 = \nabla \times \mathbf{B}^1 = \nabla \times [\nabla \times (\boldsymbol{\xi}^1 \times \mathbf{B}^0)]$. The approximation

$$\mathbf{J}^1 \simeq \nabla \times \left(-\xi_r \frac{\partial}{\partial r} \mathbf{B}^0\right) \simeq -\xi_r \frac{\partial}{\partial r}\mathbf{J}^0$$

applies in the absence of compression, $\nabla \cdot \boldsymbol{\xi} = 0$, near a mode rational surface, where $\mathbf{B}\cdot\nabla\xi$ is small and where $\xi \simeq \xi_r\hat{\mathbf{r}}$ is nearly uniform. Those kink instabilities that are localized near the surface (for example, $m \geqslant 2$ modes) are generally suppressed when J_z goes smoothly to zero at the edge of the plasma.

6.2.5

The radial magnetic field perturbation B_r^1 cannot be induced if the ridges of the perturbation are parallel to the equilibrium magnetic field in a perfectly conducting plasma $\mathbf{E}^1 \parallel \mathbf{B}^0$ in (6.2.1)). Either resistivity or a vacuum is needed.

6.3.1

We find that $v = (C_{11} \pm \sqrt{C_{12}C_{21}})/A$ for solutions of (6.3.5–6.3.6) which vary like $(r\xi_r, p^*) \sim x^v$, $x \equiv r - r_s$, in the immediate neighborhood of the radius r_s where $(\rho\gamma^2 + F^2)r \simeq Ax$. For the singularity at marginal stability, see the derivation of the Suydam criterion (6.5.11–6.5.16).

6.3.2

To show that $B_r^1 = 0$, use the integral form of Faraday's law with $\mathbf{E}^1 \cdot \mathbf{B}^0 = 0$. Resistivity allows $\mathbf{E}^1 \cdot \mathbf{B}^0 \neq 0$ and hence $B_r^1 \neq 0$.

6.4.1

By allowing $\xi \neq 0$ at the edge of the plasma, free-boundary instabilities can make more effective use of the internal plasma potential energy.

6.5.1

The Euler equation must be solved piecewise, allowing for discontinuities at singular points, as in the test function (6.5.1).

6.5.2

Shear stabilization varies like r^2 near $r=0$. The instability flattens $p(r)$ near $r=0$ and hence self-stabilizes.

6.5.3

The coefficient of $(\xi')^2$ is small and goes to zero where the eigenfunction oscillates rapidly with radius.

6.6.1

The growth rate of the internal $m=1$ mode is very similar to that of the free-boundary mode with the same mode rational surface.

6.6.2

Yes, they can be different eigenfunctions. Any given eigenfunction, however, is changed when the fixed-boundary constraint is removed from the edge of the plasma.

7 TOROIDAL INSTABILITIES

7.1.1

Substitute (7.1.5) into (7.1.3) and integrate over θ and ζ to obtain $\lambda(V,\dot\theta,\zeta)$. The variation of the magnetic energy with respect to λ involves

$$\int d^3x \tfrac{1}{2}\delta(B^2) = \int d^3x \, \mathbf{B}\cdot\mathbf{V}V \times \mathbf{V}\delta\lambda$$
$$= \int d^3x \{\mathbf{V}\cdot(\delta\lambda \mathbf{V}V \times \mathbf{B}) + \delta\lambda \mathbf{V}V \cdot \mathbf{V}\times\mathbf{B}\}.$$

The first term integrates to zero and the second term, $\mathbf{V}V\cdot\mathbf{J} = 0$, implies that no current passes through the flux surfaces in the minimum energy state.

7.1.2

First find the surfaces of constant ψ and the volume $V(R, y)$ within these surfaces by integration. The functions $\psi_{pol}(V)$, $\psi_{tor}(V)$, $I_{pol}(V)$ and $I_{tor}(V)$ can then be obtained from their definitions (4.2.1)–(4.2.4). Then use (7.1.7) and (7.1.10) in the form

$$\mathbf{B}\cdot\mathbf{V}\zeta = \dot\psi_{tor}(V) \qquad \mathbf{J}\cdot\mathbf{V}\zeta = \dot I_{tor}(V)$$
$$\mathbf{B}\cdot\mathbf{V}\theta = \dot\psi_{pol}(V) \qquad \mathbf{J}\cdot\mathbf{V}\theta = \dot I_{pol}(V)$$

and integrate along the field lines to find ζ and θ.

7.2.1

The relative magnitues of ξ^V, μ, and η would be the same, and the absolute magnitude is irrelevant in a linearized theory.

7.3.1

It seems to be a coincidence that the pressure-driven modes from the Mercier criterion have the same stability criterion as the $\mathbf{J}_{\|\mathbf{B}}$-driven kink modes in a low beta, large aspect ratio circular torus. Except for the resonance condition between the perturbation and the field lines, I am not aware of any fundamental connection.

7.3.2

The Mercier criterion indicates only that a localized eigenfunction, which is localized with infinitely many radial nodes near the mode rational surface, is stable in a toroid. A corresponding $m=2, n=1$ eigenfunction which is not so localized may still be unstable, analogous to the circular cylindrical results discussed in section 6.5.

7.3.3

The perturbation used to derive the Mercier criterion is fine-scaled only perpendicular to the flux surfaces; it is uniform along the magnetic field lines and therefore averages over the pressure gradient, curvature, and shear around the whole toroid.

7.3.4

For circular cross section $e=1$ with no diamagnetism or outward shift $Q=0$, $d=0$, the Mercier criterion reduces to $q^2 > 1$ while the Lortz sufficient criterion $1/q^2 < -1$ cannot be satisfied. In the limit of large vertical elongation $e \gg 1$, the necessary and sufficient criteria merge together. An outward shift of the magnetic axis $d > 0$ and a diamagnetic poloidal current $Q < 0$ are favorable for stability $1/q^2 < -2Q + d + \mathcal{O}(1/e)$. For fixed $d > 0$ and $Q < 0$, the elongation giving the lowest value of q in the sufficient criterion is

$$e = \left[\sqrt{\frac{3-d}{2(1-Q)}} - 1\right]^{-1/2}.$$

If $d=0$ and $Q=0$, the lowest q-value sufficient for stability near the magnetic axis is $q > 2.22$ for $e=2.1$.

7.4.1

When the system is discretized, the infinite number of resonating surfaces is replaced by a finite number, each with a resonant frequency within the range of the original continuum. How these frequencies are spaced within the continuum depends upon how the system is discretized.

7.4.2

$$\mathbf{B}_{\text{pol}}^1 = \nabla \times (\boldsymbol{\xi} \times \mathbf{B}^0) \simeq \xi \frac{nB_{\varphi 0}}{R_0} \frac{\sin(m\theta - n\varphi)}{(1 + \varepsilon\cos\theta)^2}, \quad \varepsilon \equiv a/R.$$

Hence, the $m\pm1$ harmonic is $\mathcal{O}(\varepsilon)$ of the mth harmonic.

8 HIGH BETA TOKAMAKS

8.1.1

The pressure gradient drives an instability near the outer edge of the cylindrical shell if

$$\beta_{\text{pol}} \equiv \frac{2\mu p_{\max}}{B_{z\,\text{edge}}^2} < \frac{1}{4}\frac{R}{a},$$

where a is the scale length of $p'(r) \sim p_{\max}/a$ and of $q'(r) \sim q/a$. For a more accurate estimate, a plane slab approximation can be used for the profiles within the plasma.

8.1.2

First, find the magnetic axes or points on the separatrices where $|\nabla\psi| = 0$. One separatrix looks like an ∞ with its x-point at $(0, 0)$. Its lobes surround magnetic axes at $(\pm((I_2 - \pi J_0)/2I_4)^{1/2}, 0)$. It is surrounded by another separatrix, beyond which field lines are not confined, with x-points at $(\pm((I_2 + 2\pi J_0)/8I_4)^{1/2}), (\pm((2\pi J_0 - I_2)/8I_4)^{1/2})$ provided $0 < \pi J_0 < I_2 < 2\pi J_0$ and $0 < I_4$. The vertical distance between these x-points is a measure of the width of the plasma.

8.2.1

Yes. For any perturbation that minimizes the magnetic energy by producing a vacuum magnetic field within the plasma, $\nabla \cdot \boldsymbol{\xi}$ can be made zero by adjusting the component of $\boldsymbol{\xi}$ parallel to $\mathbf{B}^0$ without further altering the perturbed magnetic field.

Comments on the Questions

8.2.2

Eq. (8.2.14) implies instability whenever

$$\beta > 1 - \frac{m - (m-q)^2 \Lambda}{q^2 + (ka)^2}, \qquad \Lambda \equiv \frac{1 + (a/r_{\text{wall}})^2}{1 - (a/r_{\text{wall}})^2}$$

so that wall and finite wavelength effects tend to stabilize higher values of beta.

8.3.1

In order to conserve toroidal flux as B_{tor} decreases within the plasma, B_{tor} must increase outside the plasma and additional current must flow in the toroidal field coils.

8.3.2

The current density is inversely proportional to the square of the distance between the membrane and the wall. Hence the total current increases like $1/(1 - (\delta/r_{\text{wall}})^2)$, where δ is the membrane shift.

8.3.3

Plasma heating raises the pressure and causes a force imbalance, resulting in an outward radial velocity and a poloidal $\mathbf{v} \times \mathbf{B}$, resulting in a negative $\frac{\partial}{\partial t} \int d\mathbf{S} \cdot \mathbf{B}_{\text{tor}} = \int d\mathbf{l}_\theta \cdot v_r \times \mathbf{B}_{\text{tor}}$, a diamagnetic poloidal current, and an additional inward $\mathbf{J}_{\text{pol}} \times \mathbf{B}_{\text{tor}}$ force to restore force balance. Similarly, the outward velocity produces a radially increasing $\mathbf{v}_r \times \mathbf{B}_{\text{pol}}$ which decreases B_{pol} and makes a smaller longitudinal current that is more spread out.

8.3.4

A force free surface current forms at the edge of the plasma. The derivative of the fluxes are not conserved within this boundary layer.

9 NONLINEAR INSTABILITY THEORY

9.2.1

Toroidal and poloidal fluxes are linked in a perfectly conducting medium, but here this seems to make no appreciable difference.

9.2.2

$$\Delta W = -\frac{1}{4\mu} B_{\theta a}^2 a^2 (r_0/a)^4.$$

9.3.1

Velocities driven by pressure gradients parallel to field lines are small since the instability has roughly the same helicity as the field lines.

9.3.2

Since the convection cells are helically twisted down the plasma column, they carry the magnetic axis and all the other field lines into helices.

9.3.3

$$\frac{\xi_a}{a} \sim \frac{13\pi(ka)^2}{3} \frac{a}{r_s}.$$

$\xi_a < r_s$ when $\left(\dfrac{r_s}{a}\right)^2 > \dfrac{13\pi(ka)^2}{3}.$

10 RESISTIVE INSTABILITIES

10.1.1

High shear is more likely to make the islands overlap for a given radial field perturbation harmonic. This is because the distance between the islands is inversely proportional to the shear while the island width is inversely proportional to the square root of the shear.

10.1.2

It is possible to have vacuum magnetic islands produced by external currents in a curved magnetic field. In this case the apparent current in the island just cancels with the current associated with the transform (10.1.1) outlined in fig. 10.2.

10.1.3

At the separatrix the field lines never make it around the magnetic island because they stagnate at the x-points. Near the axis of the island,

Comments on the Questions

the flux surfaces look like long thin cylinders with elliptical cross section with

$$q = \frac{B_{tor}k}{J_z^1}\frac{a^2+b^2}{ab}, \quad b/a = B'_{*y}/kB_x^1, \quad J_z^1 = \frac{1}{k}\frac{\partial^2}{\partial x^2}B_x^1.$$

10.1.4

Yes. Yes.

10.1.5

From $\nabla \cdot \mathbf{B} = 0$ calculate the decrease in $B_y(x)$ and the corresponding decrease in magnetic energy by $\mathcal{O}(B_y^1 B_y^0)$.

10.2.1

The conditions

$$\nabla \cdot \mathbf{v}^1 = 0, \quad \frac{\partial}{\partial z} = 0,$$

determine v_y^1.

10.2.2

When $\Delta' < 0$, the longitudinal current in the island flows in the wrong direction so that B_y^1 raises the magnetic energy rather than decreasing it.

10.2.3

The magnetic islands induced by the externally applied helical field tend to flatten the current profile and reduce the shear near the mode rational surface, and thereby reduce the growth rate of tearing modes there. They have no direct effect on Δ'.

10.4.1

Using inequality (7.2.45) together with variational form (3.3.7), it can be shown that the Rayleigh-Taylor instability can be completely stabilized by shear while the growth rate is merely diminished for the resistive interchange mode.

11 COMPARISON BETWEEN THEORY AND EXPERIMENT

11.1.1

The smallest $q=1$ radius is obtained when all the current is concentrated within the $q=1$ surface. Then $r_{q=1} = a/(q(a))^{1/2}$.

11.1.2

$c = \sqrt{2}\; r_{q=1}$.

11.2.1

At each angle, the X-ray intensity can be approximated by the length of the chord across the flat ledge of the island plus additional contributions from any part of the chord passing through the central core of the plasma. At intermediate angles the result is a sinusoidal perturbation truncated at the lower edge.

Appendix B: Glossary

Alfvén wave—a kind of plasma oscillation involving $\mathbf{B}^1 \parallel \mathbf{v}^1 \perp \mathbf{B}^0$, $\mathbf{k}$ (shear wave) with phase velocity $\omega/k = \hat{\mathbf{k}} \cdot \mathbf{B}^0/\sqrt{\mu\rho^0}$.

Aspect ratio—R/a = major radius/minor radius of toroid; = length or wavelength/$(2\pi a)$ for straight cylinder.

Ballooning instabilities—a class of instabilities where the displacement is not uniform along field lines at the mode rational surface. (sections 8.2–8.3)

Bénard convection cells—appear when a fluid is heated from below. (section 9.1)

Beta—a measure of the plasma pressure (thermodynamics energy) normalized by the magnetic field energy. (eq. 8.0.1)

Bifurcation—a point where nonuniqueness of solution develops as parameters are varied.

Critical beta—value of beta above which plasma becomes unstable, usually to ballooning modes. (sections 8.2–8.3)

Decay index—a measure of how much the vertical magnetic field decreases with major radius. (eq. 4.7.14)

Displacement vector—time integral of the perturbed velocity. (eq. 5.2.1)

Disruptive instability—abrupt expansion of the plasma column accompanied by a negative voltage spike and other effects. (sections 1.2, 11.3)

Energy principle—a variational form involving the potential energy of the linearized MHD equations. (ch. 5)

Entropy wave—convection of entropy with zero phase velocity.

Glossary

Equilibrium (MHD)—the complete balance of forces within the plasma.

Ergodic magnetic field line—comes arbitrarily close to any point on a surface or in a volume, if followed long enough.

Eulerian coordinates—are fixed in an inertial frame of reference.

Fixed-boundary of internal instabilities—grow even in the presence of a rigid, perfectly conducting boundary at the edge of the plasma.

Flute modes—are instabilities where the ridges and troughs run parallel to the magnetic field.

Flux conserving tokamak—is a tokamak heated so rapidly that the magnetic fluxes are convected along with the plasma. (section 8.3)

Flux coordinates—any coordinate system in which one of the coordinates is uniform over each flux surface. (section 7.1)

Flux surface—surface over which magnetic fluxes are uniform. (section 4.2)

Force-free field—a magnetic field for which $\mathbf{J} \times \mathbf{B} = 0$.

Free-boundary instabilities—require a vacuum region or a highly resistive region between the plasma and the wall.

Grad-Shafranov equation—for axisymmetric toroidal equilibrium. (eq. 4.4.10)

Hall effect—the $[\mathbf{J} \times \mathbf{B} - \mathbf{V}(n_e T_e)]/n_e e$ term in Ohm's law. (eq. 2.6.16)

Hamada coordinates—a particular flux coordinate system for MHD equilibrium with $\mathbf{B}$ and $\mathbf{J}$ lines straight and Jacobian = 1. (section 7.1)

High beta tokamak—has beta of order a/R or larger. (ch. 8)

Instability—the tendency to move away from an equilibrium. (section 5.0)

Interchange instability—primarily driven by pressure gradients in a curved magnetic field or a Rayleigh-Taylor type of instability.

Kink instability—primarily driven by $\mathbf{J} \parallel \mathbf{B}$.

Kruskal-Shafranov stability criterion—$q_{\text{edge}} > 1$ to stabilize $m=1$ kink. (section 6.2)

Lagrangian coordinates—move with the fluid.

$m=0$ sausage instability—local constriction of cross section. (section 6.2)

$m=1$ kink instability—plasma twists into helical corkscrew shape. (section 6.2)

Magnetic axis—field line enclosed by nested magnetic surfaces. (section 4.2)

Magnetic differential equation—of the type (7.2.39) for given σ and $\mathbf{B}^0$.

Magnetic island—a filament of field lines with their own set of nested flux surfaces and local magnetic axis. (section 10.1)

Magnetic pressure—$(1/2\mu)\nabla|B|^2$ part of the $\mathbf{J} \times \mathbf{B}$ force on plasma. (section 2.4)

Magnetic surface—is everywhere tangent to field line, same as a flux surface. (section 4.2)

Magnetic tension—$(1/\mu)B^2 \kappa$, $\kappa = \hat{\mathbf{B}} \cdot \nabla \hat{\mathbf{B}}$, part of $\mathbf{J} \times \mathbf{B}$ force on plasma. (section 2.4)

Glossary

Magnetic well—reduced toroidal field strength within plasma. (eq. 7.2.48)

Magnetoacoustic (fast) wave—compressional plasma oscillation with $\mathbf{v}^1 \parallel \mathbf{k} \perp \mathbf{B}^0$ and with phase velocity $\omega/k = \sqrt{(\Gamma p + B^2/\mu)/\rho}$.

Marginal stability—the transition from instability to stability as the equilibrium parameters are varied. (section 5.0)

Mercier stability criterion—for stability of perturbations localized arbitrarily close to the mode rational surface. (eq. 7.2.49)

MHD (magnetohydrodynamics)—a model describing the interaction between a perfectly conducting fluid and a magnetic field. (ch. 2)

Mirnov oscillations—helical magnetic field perturbation detected at the edge of tokamak plasmas. (sections 1.2 and 11.2)

Mode rational surface—magnetic surface on which field lines resonate with the helicity of a particular perturbation or instability. (section 4.2)

Necessary stability condition—violation guarantees instability.

Peeling instability—localized near edge of plasma when there is a discontinuity in the current or its derivative there. (section 6.6)

Poloidal—short way around a toroid.

Poloidal beta—measure of pressure confinement by poloidal currents. (section 4.4)

q-value—toroidal winding number/poloidal winding number. (section 4.3)

Rational magnetic surface—on which field lines close upon themselves. (section 4.2)

Rayleigh-Taylor instability—lowers the gravitational potential energy of the fluid. (ch. 3)

Resistive interchange mode—driven by pressure gradients and magnetic field curvature. (section 10.4)

Resistive tearing mode—magnetic island formation to reduce magnetic energy in regions away from the islands. (section 10.2)

Rotational transform—reciprocal of the q-value.

Sawtooth oscillations—relaxation oscillations near the center of tokamak plasmas. (sections 1.2 and 11.1)

Self adjoint—property of operator and boundary conditions satisfying, eq. (5.3.2).

Separatrix—magnetic surface separating surfaces with different topologies. (section 4.2)

Shear—spatial variation of the q-value.

Sound wave—compressional plasma oscillation with $\mathbf{v}^1 \parallel \mathbf{k} \parallel \mathbf{B}^0$ and with phase velocity $\omega/k = \sqrt{\Gamma p/\rho}$.

Spitzer resistivity—the electrical resistivity given by (eq. 2.6.18).

Stability—tendency to return to the original equilibrium given a perturbation that passes through the equilibrium state.

Sufficient stability condition—guarantees that no MHD instability will grow.

Glossary

Surface current model—all plasma currents flow on the surface only and pressure is uniform within the plasma. (section 4.6)

Surface quantity—any variable that is uniform over each flux surface. (section 4.2)

Suydam criterion—a local necessary stability criterion for straight circular cylindrical plasmas. (eq. 6.5.16)

Tokamak—a toroidal plasma confinement device. (section 1.2)

Toroid—a toroidal surface with any shape.

Toroidal—the long way around a toroid, as opposed to poloidal.

Torus—a toroidal surface with a circular cross section and a circular hole.

Variational form—an integral form of a set of eigenvalue equations. (section 3.1 and ch. 5)

Index

Adiabatic model for Rayleigh-Taylor instability, 51–54
Alfvén transit time, 202
Alfvén velocity, 28, 186
Alfvén wave, 97
Algebraic reduction of circular cylinder
  equations of motion, 109–110
Ampere's law, 28, 237
Angular momentum, conservation of, 39, 236
Aspect ratio, 10
  effect on ballooning modes, 169
  effect on high beta stability, 159–161
  finite, 10, 120, 143, 150
  for cylinder, 114, 122
  limit of large aspect ratio, 82, 114, 139–142, 159, 177
Axial contraction of elongated plasma, 152, 154
Axisymmetric stability, 84–86
Axisymmetry of tokamaks, 10

Ballooning instabilities, 23
  surface current model, 154–161
  in tokamaks, 166–172

Bénard convection cells, 178, 189, 213
Bessel functions, 111
Beta, 5, 149
  β_J, 83
  β^*, 150
  equilibrium limit, 163, 241–242
  low beta approximation, 139, 143
  in surface current model, 77, 156
  (See also Critical beta; Poloidal beta)
Bifurcation, 75, 87, 177–178, 189
Boundary conditions, 38–39, 93
  in Mercier derivation, 133
  for Rayleigh-Taylor instability, 50–51
Boundary layer
  for $m=1$ internal kink, 116–117, 178
  for tearing mode, 191, 200–203
  for resistive interchange, 211
Bubble, magnetic, 182

Charge density, 41–42
Collision time, ion-ion, 43
Compression and expansion, 29, 110.
  See also Incompressibility

Index

Computation
  linear instabilities in cylinders, 112, 116, 118–119, 124
  linear instabilities in toroids, 142–148
  simulation of $m=1$ tearing mode, 207–209
  simulation of nonlinear instabilities, 176–177, 181–189
Confinement time, 3, 7–8, 13, 15
Conservation of energy, 44, 94, 243
Conservation of momentum, 42
Conservative form of MHD equations, 37–39
Continuity. *See* Equation of continuity
Contravariant components, 126, 131
  of magnetic field, 128–129
  of current, 130
Controlled thermonuclear fusion, 3, 149–150
Convection cells
  in ballooning modes, 171–172
  instabilities in circular cylinders, 115–116
  during nonlinear evolution of instabilities, 184–187, 189
  in resistive interchange, 211
  in tearing mode, 199
Convective derivative, 28
Covariant components, 126
Critical beta, 154
  circular cylinder, 157–158, 249
  elliptical cylinder, 158
  toroid, 159, 166, 248
Cross sectional shapes, toroidal
  circular, 139, 159, 161, 172
  Doublet, 152–154
  D shape, 141, 165–166, 169–171
  elongated, 144–145, 159
Current density, 27–28, 41–42
  in Hamada coordinates, 129–130
  perturbation in kink, 244
  perturbation in tearing mode, 200, 203
Current profile
  in circular cylinder, 104, 244
  effect on elongation, 78–79, 152–154
  expansion during disruptive instability, 227–228, 230
  experimental determination of, 220
  toroidal, 67–68
  at high beta, 164–166 (*See also* Force free current)
Current, 63
  longitudinal, decreased by nonlinear kink, 180
  toroidal, limited by disruptive instability, 13–14
Curvature, 36, 97
  poloidal and toroidal components of, 139, 156
  poloidal, effect on high beta stability, 156–159
  toroidal, effect on ballooning modes, 156, 159, 167–168
  toroidal, effect on Mercier stability criterion, 139
Curvilinear coordinates, 126
Cut surfaces (for magnetic fluxes), 62

Decay index, 84–85
Density, 41
  limited by disruptive instability, 14
Diagnostics, experimental, 8, 12–13, 16–19, 25–26, 220, 224, 227
Diamagnetic currents, 165–166
Diamagnetic frequency, 206
Diamagnetism, 68–69
  contributes to force along major radius in tokamaks, 80–81, 83
Displacement current, 28, 237
Displacement vector, 91–92
Disruptive instability, 11–14, 226–231
Double adiabatic model, 42
Doublet shape, 152–154
Dynamic stabilization, 56–58, 239
Dynamic state, 28, 40, 92

Eigenfunctions, 91, 95
Eigenvalue equations, 94–95, 98
  circular cylinder, 109–112
Eigenvalue method, computational, 142–143
Eigenvalues, see spectrum
Electric field
  different frames of reference, 28–29, 31
  boundary conditions, 38–39
Electrical permittivity, 4
Electromotive force, 234
Electrostatic force, 32, 42
Elongated cross section
  equilibrium, 71–79

Index

Elongated cross section (continued)
 experimental, 151–154
 linear stability, 158–163
 nonlinear kink, 177
Energy principle, 93–101
 circular cylinder, 112–114
 normalization of, 98–99
 surface current model, 155–157
 (See also Stability conditions (analytic) for applications of the energy principle)
Entropy, 39
 and Rayleigh-Taylor instability, 52–54
Entropy wave, 54
Equation of continuity, 27, 37, 41, 49
Equilibrium, MHD, 59–60, 87–88
 beta limit, 163, 241–242
 circular cylinder, 103–104
 elongated cylinder, 71–79, 153, 240–241
 flux-conserving sequences, 163–166
 Hamada coordinates, 129, 131
 surface current model, 76–77
 tokamak, 79–86 (See also Bifurcation, and Grad-Shafranov equation)
Ergodic field lines, 60–62, 227
Euler equations, 99, 238
 circular cylinder, 115, 120, 245
Eulerian coordinates, 37–38, 91–92
Expansion, during disruptive instability, 13, 228–230
Experimental devices, world survey by IAEA, 24
Experimental observations, 4–26, 217–232

Faraday's law, 27–28, 30–31, 37–38
Finite element method, 142–143
Flow pattern. See Vortex cells
Flux, magnetic, 30, 62
 at boundary, 38–39
 helical, 196
Flux conserving tokamak, 163–166, 174, 249
Flux coordinates, 125, 127–130, 147, 246
Flux surfaces, 63
 elongated, 71–79, 153
 Doublet, 153
 shifted, 139, 141
 toroidal, 165
 triangularity of, 141

Field reversal, 189. See also reversed field pinches
Force free current or field, 69, 249
 effect on high-beta stability, 150, 169
Fourier analysis, 49, 55, 109, 160

Galilean invariance and Faraday's law, 29, 31
Gauss's theorem, 30, 38, 60, 155
Glasser, Greene, and Johnson dispersion relation, 212
Grad-Shafranov equation, 66–68, 70, 87, 164
Gravitational acceleration
 and the Rayleigh-Taylor instability, 48, 238
Green function, 74
Growth rate
 ballooning, 169
 circular cylinder, internal instabilities, 118
 elongated rectangular cylinder, internal instabilities, 162
 $m=1$ free boundary, 122
 $m=1$ internal, 117–118
 $m=1$ tearing, 205–206
 $m \geqslant 2$ tearing, 202
 Rayleigh-Taylor, 50–54, 56, 237–239
 resistive interchange, 211–213
 toroidal, 145–146
Gyro frequency, 43–44

Hamada coordinates, 129–132
Hall effect, 43–44, 225
Harmonic mixing in toroidal geometry, 143–146, 171
Heat conductivity, 44, 223
Heating, plasma, 4, 11, 218–221
Helical magnetic field, applied, 202, 226–227
Helical plasma distortion, observed, 11–13, 229
Helical flux function, 196
Helical pitch of magnetic field line, 65, 140
Helically symmetric equilibrium, 70
Helically symmetric perturbation, nonlinear evolution of, 177, 179–182
High beta
 conferences, 173
 effect on kink modes, 182–183

High beta (continued)
(*See also* Ballooning modes and high beta tokamak)

IAEA conferences, 24
Impurities, 9, 13, 15, 18
Incompressibility
 ballooning modes, 155, 167, 248
 circular cylinder instabilities, 113
 Mercier stability derivation, 132–135
 effect on Rayleigh-Taylor instability, 48–54
 tearing modes, 199, 251
Indicial equation, 110, 120
Inductance, 82–83, 166
Inequality used for Suydam or Mercier derivation, 138
Initial value method, 98, 142
Instability, 89–90, 94
 axisymmetric toroidal, 84–85
 ballooning, 154, 166–172, 174
 current driven, 104–109
 disruptive, 11–15, 25, 226–231
 effect of elongated cross section, 158–163
 fixed boundary or internal, 114–121, 144, 162–163, 183–188
 free boundary, 113, 121–123, 142, 145, 154–161, 179–183
 effect of high beta, 154–163, 166–172
 interchange, 102 (*See also* Rayleigh-Taylor; Suydam; Mercier; Resistive interchange; Localized instabilities)
 kink, 102 (*See also* $m=1$ kink; $m \geqslant 2$ kink)
 localized, 8, 49, 117–120, 132, 211
 $m=0$ sausage, 5–6, 25, 104–105, 244
 $m=1$ internal kink, 114–117, 183–186, 206
 $m=1$ kink, 6–7, 25, 105–107, 121–122, 179–182, 231–232
 $m \geqslant 2$ internal, 117, 186–188
 $m \geqslant 2$ kink, 108–109, 182–183, 122
 $m \geqslant 2$ resistive tearing mode, 198–205
 Mercier, 130–141
 peeling, 122
 Rayleigh-Taylor, 47–58, 237–239
 resistive interchange, 209–213
 Suydam, 120–121
 toroidal, 142–146, 213
Islands. *See* Magnetic islands
Isothermal model, 54

Jacobian, 126, 129
$\mathbf{J} \times \mathbf{B}$ force, 28, 31–32, 36–37, 60, 105

KeV, 3
Kink mode, 102. *See also* Instability; Stability conditions
Kruskal-Shafranov stability criterion, 7, 10–11, 107–108, 122

Lagrange multiplier, 98, 137
Lagrangian coordinates, 39, 91–92
Large aspect ratio expansion, 82, 114, 139–142, 159, 177
Larmor radius stabilization, 206, 232
Lawson criterion, 3, 26
Limiter in tokamaks, 9
Linearized MHD equations, 90–93, 109–112
Linear stability. *See* Stability
Lorentz force, 31
Lortz and Nührenberg stability criterion, 140–141

Magnetic axis, 33, 61, 65, 240
 of an island, 192, 197
 evaluation of Mercier criterion, 139–141
Magnetic differential equation, 136–137
Magnetic energy, 15, 56, 128, 234, 236, 251
Magnetic field
 divergence-free property, 29–30
 in flux coordinates, 127–129
 being frozen into a perfectly conducting fluid, 32–35, 164
 poloidal and toroidal in tokamak, 9–10
Magnetic field lines, 29, 32–35, 45, 235
 ergodicity, 29, 227
Magnetic field perturbation
 in ballooning modes, 167
 Δ' for tearing modes, 200, 202, 204, 212, 251
 in linearized MHD equations, 92
 in magnetic island, 193, 194–197
 in Mercier stability derivation, 132–136

Index

Magnetic field perturbation (continued)
  during Mirnov oscillations and disruptive instability, 16, 226, 229
  during nonlinear evolution of internal instability, 183–187
  in toroidal instabilities, 143–144, 248
  in vacuum, 111–112, 155
Magnetic flux. See Flux
Magnetic islands, 61, 192–197, 214, 250
  contributing to the disruptive instability, 227–230
  Mirnov oscillations, 223–224
  $m=1$ resistive tearing mode, 207–208
  $m \geqslant 2$ resistive tearing mode, 198–199, 203–204
  width of, 195–197, 204
Magnetic permeability (vacuum), 4, 28
Magnetic pressure, 36, 42, 233, 236
  in equilibrium, 60, 72
  in the surface current model, 77
Magnetic Reynold's number, 202
Magnetic surface, 60. See also Flux surfaces
Magnetic tension, 36, 236
  in equilibrium, 60, 71
  effect on Rayleigh-Taylor instability, 56
Magnetic well, 138, 140
Magnetoacoustic wave, 97–98
Magnetohydrodynamic (MHD) model, 27–29, 40–44
Marginal stability, 89–90, 95, 99, 242–243
Maxwell stress tensor, 37, 60
Mercier stability criterion, 138
  applications, 139–141, 247
  derivation, 130–138, 146–147
Metric tensor, 126
Mirnov oscillations, 15–20, 25, 223–226, 231
  effect on disruptive instability, 227
Mode numbers, 13, 117–119, 132, 161
  in toroid, 144
  nonlinear coupling, 177
Mode rational surface, 61
  in cylinder, 111, 116–117
  in Mercier stability derivation, 132, 134–135
  magnetic islands, 193
Moment equations, 40–45
Momentum equation (2.1.1), 27
  at boundary, 39
  conservative form, 37

Necessary conditions for stability, 99, 130
Negative voltage spike, 11–13, 230
Neoclassical transport theory, 40
Neutron production from sausage instability, 25
Nonlinear evolution
  free boundary kink, 179–183
  fixed boundary instabilities, 183–188
  Rayleigh-Taylor instability, 51
Nonlinear saturation
  free boundary kink, 180–183
  internal $m=1$, 188, 250

Octopole field, 152
Ohmic heating, 4, 151
Ohm's law, 43
  ideal, 31
Oscilloscope traces, 18, 222
Overstability, 90, 213

Pendulum
  analogue of Rayleigh-Taylor instability, 56–57, 239
  in magnetic field, 35
Perturbed magnetic field or pressure. See Magnetic field or pressure
Plane slab, 48–56, 195–196
Plasma, 28
Plasma physics, 24
Poloidal beta, 68–69, 141, 149, 169
Poloidal field coils, 79
Potential energy, 93. See also Energy principle
Poynting flux, 38, 237
Pressure (plasma), 28, 41–42, 63
  effect on ballooning modes, 167–168
  contributes to force along major radius in tokamaks, 79–80
Pressure perturbation, 92, 144, 161, 170, 243
Pressure profile, 67, 170
  nonlinear evolution of, 184–187
Principle of least action, 94
Profiles. See Current; Pressure; Temperature

Index

q-value, 64–66, 129, 239–240
  in axisymmetric toroid, 164, 220
  in circular cylinder, 103–104, 107, 113–114, 244
  convected by perfectly conducting fluid, 164–165
  in elongated cylinder, 73
  within magnetic islands, 197, 251
  profile of, 219, 223, 244, 252
  in tokamak experiments, 13–14, 17, 219
  in stability diagrams, 118, 144–145, 157–160, 168–169

Radial nodes, 112, 117, 119–121
Rational magnetic surface, 61. *See also* Mode rational surface
Ratio of specific heats, 29
Rayleigh-Taylor instability, 47–58, 237–239
Reconnection of magnetic field lines, 207, 209, 214–215, 228
Resistive interchange mode, 209–213, 215–216
Resistive skin time, 202
Resistive tearing mode, 198–209. *See also* Tearing mode
Resistivity
  needed for magnetic island formation, 195
  Spitzer (classical), 43
Reversed field pinches, 124, 186
Rotation of Mirnov oscillations, 225–226
Rotational transform, 66
Runaway electrons, 9, 11, 14, 228

Safety factor. *See* q-value
Saturation. *See* Nonlinear
Sausage instability, 5–6, 25, 104–105
Sawtooth oscillations, 20–21, 177, 218–223, 227, 230–231
Scalar potential, 155
Screw pinch, 102
Self-adjointness, 93, 95
Separatrix, 61
  in elongated equilibria, 72, 248
  around magnetic island, 194–196, 250
  in toroidal equilibrium, 163
Shafranov-Solov'ev equilibrium, 71, 240
Shafranov-Yurchenko stability criterion, 139–140

Shear, 54
  effect on ballooning modes, 169
  in circular cylinder, 104, 114–117
  global, 136, 138
  effect on interchange modes, 211, 213
  effect on $m=1$ internal modes, 206
  effect on Mercier stability, 139
  effect on nonlinear kink, 182
  effect on Rayleigh-Taylor instability, 56
Shell, conducting, in tokamaks, 10, 79
Shock heated experiments, 151–154
Shooting codes, 112
Singular perturbation analysis
  linear $m=1$, 115–117
  nonlinear, 178, 189
  resistive instabilities, 191
Soft X-ray signal, 17–21, 25, 220–224, 227–228, 252
Spectrum of eigenvalues, 49
  circular cylinder, 112, 123
  computation of, 99, 143
  Rayleigh-Taylor instability, 49–51, 54
Stability, 89, 94, 242–243
Stability conditions (analytic) for
  axisymmetric stability, 85
  ballooning modes, 166
  circular cylinder, high beta surface current model, 157, 249
  elliptical cylinder, $m \geqslant 1$ kink modes, 160
  $m=0$ sausage, 244
  $m=1$ free-boundary kink, 107, 122
  $m \geqslant 2$ free-boundary kink, 109
  Mercier criterion, 138–141
  Rayleigh-Taylor, 51–54, 56, 237–239
  resistive interchange, slab, 211, toroidal, 212
  resistive tearing, $m=1$, 205–206, 218
  $m \geqslant 2$, 202, 213, 251
  Suydam criterion, 120, 152, 248
Stability diagrams (computed or experimental) for
  ballooning modes, 152–162, 168–169
  disruptive instability, 14
  fixed-boundary modes in circular cylinder, 118

Index

Stability diagrams (continued)
  fixed-boundary modes in rectangular cylinder, 162
  fixed-boundary modes in toroid, 144
  free-boundary modes in toroid, 145
  free-boundary modes in surface-current finite-beta configurations, 157–161
  sawtooth oscillations, 20
Stable oscillations, 95, 111, 119, 142
State. See Dynamic state
Stream function, 66–67
  for arbitrary magnetic field, 127–128
  circular cylinder equilibrium, 103
  for equilibria with continuous symmetry, 70
Sturm-Liouville problem, 112
Sufficient conditions for stability, 99, 130
Superthermal ions accelerated by instabilities, 5, 230
Surface quantities, 62–63, 67
Surface current model, 76–77
  energy principle for, 96, 113
  finite beta stability, 154–160
  nonlinear kink, 179–180
Suydam stability criterion, 120–121
  applied to cylindrical slab, 151–152

Tearing mode, resistive, 198
  $m = 1$, 205–209, 215
  $m = 1$, effect on sawtooth oscillations, 218, 221–222
  $m \geqslant 2$, linear theory, 198–202, 215
  $m \geqslant 2$, nonlinear theory, 203–204, 215
  $m \geqslant 2$, effect on Mirnov oscillations, 224–225
  $m \geqslant 2$, effect on disruptive instability, 227–228
Temperature, 41
  profile during sawtooth oscillation, 218–219
  profile during disruptive instability, 228
Test functions, 50, 98
  $m = 1$, 114
  Mercier, 137
  Suydam, 120
Theta pinch, 102
Tokamaks, 8–11, 24
  equilibrium, 79–86
  stability, 11–21, 218–231
Torsion, 97
Turbulence, 178–179, 189

Vacuum region, 39
  circular cylinder, 109, 111–113
  extension of energy principle, 96
Variational form, 50–56
  in Mercier derivation, 137
Velocity, fluid, 41, 225. See also Vortex cells
Vertical field, 79, 83, 87–88
  produces separatrix, 163
Viscosity, 42–43
Von Hagenow S curve, 78–79
Vortex cells in
  ballooning mode, 171–172
  circular cylinder instabilities, 115–116
  internal instabilities in rectangular cylinder, 161, 163, 183–187
  toroidal instabilities, 143–146

Wall
  rigid, perfectly conducting, as a boundary condition, 39
  stabilization, 122, 182
  tokamak, 9

X rays
  hard, 13, 17, 228, 252
  soft, 17–21, 25, 220–224, 227–228, 252

Z-pinch, 102